a mathematical
theory of evidence

a mathematical theory of evidence

Glenn Shafer

Princeton University Press

Published by Princeton University Press, Princeton and London

Library of Congress Cataloging in Publication Data will
be found on the last printed page of this book

This book has been composed in Varityper Bookman
Printed in the United States of America
by Princeton University Press, Princeton, New Jersey

This book is dedicated to the memory
of Tempa Shafer and Ralph Fox,
two of my teachers.

FOREWORD

Glenn Shafer has greatly extended, refined, and recast a theory on which I expended much effort in the 1960's. The mathematical theory has been rebuilt around the concept of combining simple support functions and their corresponding weights of evidence, and many interesting new mathematical results are presented. Simultaneously, the theory has broken free of the narrow statistical confines of random sampling in which I worked, and has been reexpressed in terms of more general relevance such as belief, support, and evidence.

My own work grew out of attempts first to understand and then to replace the fiducial argument of R. A. Fisher. Beginning about fifty years ago, the British-American school believed that it had, by various pragmatic devices, banished the Bayesian scourge. But I gradually came to the perception that my revision of the fiducial argument was merely a loosening of the formalism of Bayesian inference, almost identical in purpose, yet more cautious because it assigned propositions lower probabilities (Shafer's degrees of belief) rather than simple additive probabilities.

I differ from Shafer in that I am comfortable with the view that subjective, epistemic probability is the essential concept, while chance or physical probability is only a subspecies which scientific tradition has come to regard as "objective." Moreover, I believe that Bayesian inference will always be a basic tool for practical everyday statistics, if only because questions must be answered and decisions must be taken, so that a statistician must always stand ready to upgrade his vaguer forms of belief into precisely additive probabilities. It is nevertheless very important to study theories which permit discrimination of circumstances where knowledge is secure enough to permit fair bets from other circumstances where the concept of fair bet becomes increasingly meaningless. One principle of rationality would be to confine one's enterprises to the former type of circumstance.

The mysteries of probable reasoning are unlikely ever to disappear, but the techniques, models, and formalisms in current use will undergo gradual sideways motion as well as growth. One current trend is to worry more about the prior knowledge embodied in the specification (i.e., assumed parametric model) part of the traditional models of statistics, rather than focusing entirely on the "prior" distributions of Bayesian inference. Witness the current interest in robustness studies. Shafer points out that the data represented by a specification cannot offer support for that specification, and he offers weight of conflict as a new criterion for comparing specifications. As one thought for the future, I wish to suggest that, just as traditional Bayesian reasoning has been shaken loose from its moorings, perhaps there will appear a comparable weakening of the strong form of information implied by a typical specification.

A. P. DEMPSTER

JULY 1975

PREFACE

In the spring of 1971 I attended a course on statistical inference taught by Arthur Dempster at Harvard. In the fall of that same year Geoffrey Watson suggested I give a talk expositing Dempster's work on upper and lower probabilities to the Department of Statistics at Princeton. This essay is one of the results of the ensuing effort. It offers a reinterpretation of Dempster's work, a reinterpretation that identifies his "lower probabilities" as epistemic probabilities or degrees of belief, takes the rule for combining such degrees of belief as fundamental, and abandons the idea that they arise as lower bounds over classes of Bayesian probabilities.

In writing the essay I have tried to combine mathematical rigor with an emphasis on the intuitive ideas that the mathematical definitions represent. I have provided thorough proofs for all displayed theorems, but in order not to clutter the reader's view of the main ideas, I have banished these proofs to appendices at the ends of the chapters. And in an effort to keep the mathematics manageable for both author and reader I have limited the exposition to the case where the set of possibilities one considers is finite. Thus even the proofs, though they are sometimes closely reasoned, never appeal to mathematical facts more advanced that the binomial theorem or the properties of the exponential function. This de-emphasis of the purely mathematical aspects of the essay's theory will not, I hope, deter readers from attacking the genuine mathematical challenges involved in generalizing the theory to infinite sets of possibilities and in extending it in the direction suggested in Chapter 8.

*　　　　　*

I am indebted to Art Dempster personally as well as intellectually, and I would like to thank him for personally attending to my work and for consenting to write a foreword to this essay.

I would also like to thank the many friends, colleagues, and teachers who have helped me with these ideas during the past several years. Foremost among these is my wife Terry, whose comments have been helpful at every stage. The early encouragement of Peter Bloomfield, Gary Simon, and Geoffrey Watson was crucial, and at the final stage of preparing the exposition I profited particularly from comments by Frank Anscombe, Bill Homer, and Richard Jeffrey. Larry Rafsky of Bell Telephone Laboratories checked most of the proofs.

I am indebted to Princeton University and the taxpayers of the United States for financial support. I learned most of the basic ideas of this essay during the academic years 1971-73, while I was supported by a National Science Foundation graduate fellowship, and I learned more during the summer of 1973, while my salary was paid by contract N00014-67A0151-0017 from the Office of Naval Research. I began writing the essay in the spring of 1974 while teaching an undergraduate course called *Probability and Scientific Inference*, and I completed the bulk of it during the summer of 1974, with financial support from National Science Foundation grant GP-43248.

Finally, I would like to thank all those involved in the physical preparation of the book. Florence Armstrong did an excellent and timely job of typing the draft originally submitted to the Princeton University Press. And special commendation is due the dedicated individuals associated with the Press itself who make it possible to publish this and other scholarly books in this relatively inexpensive form.

GLENN SHAFER
PRINCETON, NEW JERSEY
JUNE 26, 1975

CONTENTS

a mathematical
theory of evidence

A MATHEMATICAL THEORY OF EVIDENCE

Glenn Shafer

CHAPTER 1. INTRODUCTION

By *chance* I mean the same as probability.
THOMAS BAYES (1702-1761)

Ainsi, un événement aura, par sa nature, une chance plus ou moins grande, connue ou inconnue; et sa probabilité sera relative à nos connaissances, en ce qui le concerne.
SIMEON DENIS-POISSON (1781-1840)

The mathematical theory presented in this essay is at once a theory of evidence and a theory of probable reasoning. It is a theory of evidence because it deals with weights of evidence and with numerical degrees of support based on evidence. It is a theory of probable reasoning because it focuses on the fundamental operation of probable reasoning: the combination of evidence.

The theory begins with the familiar idea of using a number between zero and one to indicate the degree of support a body of evidence provides for a proposition – i.e., the degree of belief one should accord the proposition on the basis of the evidence. But unlike past attempts to develop this idea, the theory does not focus on the act of judgment by which such a number is determined. It focuses instead on something more amenable to mathematical analysis: the combination of degrees of belief or support based on one body of evidence with those based on an entirely distinct body of evidence. The heart of the theory is Dempster's rule for effecting this combination.

This introductory chapter provides a synopsis of our theory, contrasts its subject matter with the subject matter of the mathematical theory of chance, and contrasts its approach with the approach of a more familiar theory of partial belief: the Bayesian theory.

The synopsis follows in §1 below; it describes, of course, only the most salient features of the following chapters. The theory of chance is taken up in §§2-4, where we learn how chance differs from partial belief and why chances are sometimes interpreted as degrees of belief. The Bayesian theory is introduced in §5, where it is explained as an attempt to appropriate the rules for chances as rules for degrees of belief. And the Bayesian theory is contrasted with and related to our theory in §§6-11. Finally, §12 takes up the meaning of the word "probability," a semantic issue that often entangles discussions of chance and partial belief.

I discuss the Bayesian theory at length in this chapter because of the essential role this very controversial theory has played in the historical development of the idea of partial belief or "subjective probability." In the past almost all students of this idea have tied it to the Bayesian theory: those who have been committed to the value of the idea have invariably adopted and defended the Bayesian theory, while those who have rejected the Bayesian theory have tended to consider their objections to the theory proof of the inviability of the idea. As we will see in §§5-11 below, our theory frees the idea of partial belief from the Bayesian theory and develops it in a way that should appeal to both sides in the Bayesian controversy. As I explain in §5 and §§10-11, our theory includes the Bayesian theory as a special case and thus retains at least some of the attraction of that theory. And as I explain in §§6-9, the divergences between our theory and the Bayesian theory are closely related to the objections that the opponents of Bayesian theory have found so convincing.

§1. Synopsis*

Suppose Θ is a finite set, and let 2^{Θ} denote the set of all subsets

*For an explanation of the mathematical notation, see the appendix to Chapter 2.

of Θ. Suppose the function $\mathrm{Bel}: 2^{\Theta} \to [0,1]$ satisfies the following conditions:

(1) $\mathrm{Bel}(\emptyset) = 0$.

(2) $\mathrm{Bel}(\Theta) = 1$.

(3) For every positive integer n and every collection $A_1, \cdots, A_n$ of subsets of Θ,

$$\mathrm{Bel}(A_1 \cup \cdots \cup A_n) \geq \sum_i \mathrm{Bel}(A_i) - \sum_{i<j} \mathrm{Bel}(A_i \cap A_j) + - \cdots + (-1)^{n+1} \mathrm{Bel}(A_1 \cap \cdots \cap A_n).$$

Then Bel is called a *belief function* over Θ. This essay explores the possibility of using such functions to represent partial belief.

Such a possibility arises when the set Θ is interpreted as a set of possibilities, exactly one of which corresponds to the truth. For each subset A of Θ, the number Bel(A) can then be interpreted as one's degree of belief that the truth lies in A. And rules (1)-(3) can be understood as rules governing these degrees of belief.

EXAMPLE 1.1. *The Ming Vase.* I contemplate a vase that has been represented as a product of the Ming dynasty. Is it genuine or is it counterfeit?

Let θ_1 correspond to the possibility the vase is genuine, θ_2 to the possibility it is counterfeit. Then

$$\Theta = \{\theta_1, \theta_2\}$$

is the set of possibilities, and

$$2^{\Theta} = \{\emptyset, \Theta, \{\theta_1\}, \{\theta_2\}\}$$

is the set of its subsets. A belief function Bel over Θ represents my belief if $\mathrm{Bel}(\{\theta_1\})$ is my degree of belief that the vase is genuine and $\mathrm{Bel}(\{\theta_2\})$ is my degree of belief that it is counterfeit.

Denote $s_1 = \mathrm{Bel}(\{\theta_1\})$ and $s_2 = \mathrm{Bel}(\{\theta_2\})$. Then rule (3) above imposes certain restrictions on the values that s_1 and s_2

can take. But these restrictions are not severe; as we will see in Chapter 2, they boil down to the requirement that $s_1 + s_2 \leq 1$. Notice that the different pairs of values (s_1, s_2) satisfying this requirement correspond intuitively to different situations with respect to the weight of evidence on the two sides of the issue. If I have little evidence on either side – little reason either to believe or disbelieve the genuineness of the vase – then I will set both s_1 and s_2 very low; in the extreme case of no evidence at all, I will set both exactly equal to zero. If, on the other hand, the evidence almost conclusively favors the genuineness of the vase, then I will set s_1 near one and s_2 near zero. Finally, substantial evidence on both sides of the issue will lead me to profess some belief on both sides; I might, for example, set $s_1 = .4$ and $s_2 = .3$. ■

There would be no sense in any claim that degrees of belief are compelled to obey rules (1)-(3). And I do not pretend that an individual would be "irrational" to profess degrees of belief that do not obey these rules. But the rules are intuitively attractive and essential to our theory. They are intuitively attractive because they derive from a simple picture – a picture in which one's belief is divisible and having a certain degree of belief amounts to committing a certain portion of one's belief. And they are essential to our theory because only those set functions that obey them can be combined by Dempster's rule of combination.

Mathematically, Dempster's rule is simply a rule for computing, from two or more belief functions over the same set Θ, a new belief function called their *orthogonal sum*. The burden of our theory is that this rule corresponds to the pooling of evidence: if the belief functions being combined are based on entirely distinct bodies of evidence and the set Θ discerns the relevant interaction between those bodies of evidence, then the orthogonal sum gives degrees of belief that are appropriate on the basis of the combined evidence.

Our exposition begins with a study of belief functions in Chapter 2 and with a presentation of Dempster's rule in Chapter 3. It then turns, in Chapter 4, to the task of illustrating Dempster's rule and developing the perspective the rule affords on the representation of evidence. The first step in developing this perspective is to identify the simple support functions.

A belief function $\mathrm{Bel}: 2^{\Theta} \to [0,1]$ is called a *simple support function* if there exists a non-empty subset A of Θ and a number s, $0 \leq s \leq 1$, such that

$$\mathrm{Bel}(B) = \begin{cases} 0 & \text{if } B \text{ does not contain } A \\ s & \text{if } B \text{ contains } A \text{ but } B \neq \Theta \\ 1 & \text{if } B = \Theta. \end{cases}$$

Such a belief function corresponds intuitively to a body of evidence whose precise and full effect is to support the subset A to the degree s. By virtue of supporting A, such evidence also supports any subset containing A. But it provides no support for the subsets of Θ that do not contain A.

When combined by Dempster's rule, two simple support functions focused on the same subset A yield another simple support function focused on A. But when we combine two or more simple support functions with different foci, we typically obtain a belief function that is not a simple support function, but is rather more complicated. Combination thus leads us from the class of simple support functions to a larger class of belief functions, a class I call the *separable support functions*.

As the nomenclature suggests, the simple and separable support functions are included in a yet larger class of belief functions called *support functions*. The support functions are easy to describe: they include all those belief functions that can be obtained by beginning with a separable support function on a certain set of possibilities and then "coarsening" that set of possibilities by neglecting to distinguish between certain of its elements. But a precise statement of what is meant by coarsening requires some care, and our study of support functions thus follows a thorough discussion, in Chapter 6, of coarsening and the opposite process of refining.

Closely related to the idea of a simple support function is a second idea also introduced in Chapter 4 – the idea of a *weight of evidence*. The degree of support s that a simple support function assigns its focus A should, intuitively, be determined by the weight w of the evidence pointing to A; s should increase as w increases, tending towards its maximum value 1 as w becomes indefinitely large. By requiring that the combination of two simple support functions on A should correspond to addition of the corresponding weights of evidence, we discover, in Chapter 4, a more precise relation:

$$s = 1 - e^{-w} .$$

Thus we may call

$$w = -\log(1-s)$$

the weight of evidence associated with the simple support function.

The description of simple support functions in terms of weights of evidence extends straightforwardly to separable support functions; as we see in Chapter 5, a separable support function over a set Θ corresponds to the specification of a weight of evidence for each proper non-empty subset of Θ. It is more problematic, and more interesting, to try to extend the idea of weights of evidence to the whole class of support functions. Indeed, since a given support function over a set Θ can always be obtained from any of many different separable support functions over various refinements of Θ, what can be said about the weights of evidence underlying it? I have not solved the mathematical problem posed by this question. But I do propose a conjecture, the "weight-of-conflict conjecture," which may offer the basis for a solution. The conjecture is stated in §6 of Chapter 5, and its implications are discussed in §§3-4 of Chapter 8.

The idea of weights of evidence also helps us understand the nature of the *quasi support functions* – the belief functions that do not qualify as support functions. As we see in Chapter 9, a quasi support function can always be obtained as the limit of a sequence of support functions. And roughly speaking, a sequence of support functions that has a quasi support function as its limit always exhibits contradictory weights of evidence tending simultaneously to infinity; the values of the quasi support function are determined by the finite values to which the differences among these contradictory weights tend. It is the intuitive dubiety of this picture of finite differences among infinite weights that motivates the name *support function* for those belief functions that can be obtained from weights of evidence without a limiting process and the name *quasi support function* for those that can be obtained only through such a limiting process.

The demonstration in Chapter 9 of the nature of quasi support functions marks the end of the essay's central thread of mathematical argument. Following this demonstration I introduce consonant support functions (Chapter 10), and discuss the problem of statistical inference (Chapter 11 and part of Chapter 12). And finally, in Chapter 12, I point out the role of assumption in the formulation of one's set of possibilities Θ and emphasize the consequent limitations on our theory.

§2. The Idea of Chance

For several centuries, the idea of numerical degree of belief has been identified, in both popular and scholarly thought, with the idea of chance. For most laymen and even many mathematicians, the two ideas are united under the name *probability*. But the reader will find the present essay intelligible only if he rejects this unification. Both numerical degrees of belief and chances have their roles to play in the following chapters, but most of the numerical degrees of belief studied there are not chances and do not obey all the rules obeyed by chances.

Chances arise only when one describes an *aleatory* (or *random*) *experiment*, like the throw of a die or the toss of a coin. The outcome of such an experiment varies randomly from one physically independent trial to another, and the proportion of the time that a particular one of the possible outcomes tends to occur is called the chance of that outcome. If $\mathcal{X}$ denotes the set of all possible outcomes, and if one specifies the chance $q(x)$ for each possible outcome $x \in \mathcal{X}$, then one has specified a function

$q : \mathcal{X} \to [0,1]$. If the set $\mathcal{X}$ is finite – and I shall assume throughout this essay that it is – then the function q completely specifies the chances involved in the experiment. I will call q the *chance density* governing the experiment.

In addition to satisfying $0 \leq q(x) \leq 1$ for all $x \in \mathcal{X}$, a chance density must also satisfy

$$\sum_{x \in \mathcal{X}} q(x) = 1 ; \qquad (1.1)$$

being proportions, the chances must add to one. But this is the only condition that a function $q : \mathcal{X} \to [0,1]$ must satisfy in order to qualify as a chance density. (In particular, $q(x)$ may be zero for some x; such an x will not be "possible" after all.) There are, therefore, many chance densities on a given set $\mathcal{X}$, and knowledge of the set $\mathcal{X}$ of possible outcomes of an aleatory experiment hardly tells us what chance density governs that experiment.

EXAMPLE 1.2. *Dime-Store Dice.* Willard H. Longcor of Waukegan, Illinois, reported in the late 1960's that he had thrown a certain type of inexpensive plastic die over one million times, using a new die every 20,000 throws.* In order to avoid recording errors, Longcor recorded only whether the outcome of each throw was odd or even, but a group of Harvard scholars who analyzed Longcor's data and studied the effects of the drilled pips in the die guessed that the chances of the six different outcomes might be approximated by the numbers in the following table:

x	1	2	3	4	5	6	Total
q(x)	.155	.159	.164	.169	.174	.179	1.000

*See the article by Iverson, et al., *Psychometrika*, 1971.

They obtained these numbers by calculating the excess of even over odd in Longcor's data and supposing that each side of the die is favored in proportion to the extent that it has more drilled pips than the opposite side. The 6, since it is opposite the 1, is the most favored. ■

Besides the proportion of the time that the actual outcome of an experiment tends to be a particular element x of $\mathcal{X}$, we may also interest ourselves in the proportion of the time that the actual outcome tends to be in a particular subset U of $\mathcal{X}$. This latter proportion is called the chance of U occurring; it may be denoted by Ch(U) and calculated by adding the chances for the various elements of U:

$$\mathrm{Ch}(U) = \sum_{x \in U} q(x) . \tag{1.2}$$

The function Ch obviously conveys exactly the same information as q. I will call it the *chance function* corresponding to q.

A function $\mathrm{Ch} : 2^{\mathcal{X}} \to [0,1]$ is a chance function – i.e., it can be obtained from some chance density q on $\mathcal{X}$ – if and only if it obeys the following rules:

(1) $\mathrm{Ch}(\emptyset) = 0$.

(2) $\mathrm{Ch}(\mathcal{X}) = 1$.

(3) If $U, V \subset \mathcal{X}$ and $U \cap V = \emptyset$, then $\mathrm{Ch}(U \cup V) = \mathrm{Ch}(U) + \mathrm{Ch}(V)$.

(This is proven in Chapter 2 below – see Theorem 2.9. Notice that rule (1), which assigns zero to the actual outcome being in the empty set, accords with (1.2) by means of the mathematical convention that a sum of no terms is zero.) These three rules may be called the *basic rules for chances*. The third one, which says that the chances of disjoint sets add, is called the *rule of additivity* for chances.

As a comparison of the basic rules for chances with the rules for belief functions (§1 above) will reveal, it is the rule of additivity that I reject as a rule for degrees of belief.

§3. The Doctrine of Chances

The theory of chance – the mathematical theory that Abraham De Moivre (1667-1754) called the doctrine of chances – begins with the simple idea of a chance density $q : \mathcal{X} \to [0, 1]$ describing an aleatory experiment with outcomes in $\mathcal{X}$. But its fascination and vigor derive from the idea of a *product* chance density and from the related but less important idea of a *conditional* chance density.

The idea of a product chance density arises when one contemplates a sequence of n physically independent trials of an experiment governed by a given chance density $q : \mathcal{X} \to [0, 1]$. Such a sequence always results, of course, in a sequence of n elements of $\mathcal{X}$ – i.e., it results in an element $(x_1, \cdots, x_n)$ of the Cartesian product $\mathcal{X}^n$. One may therefore think of the n physically independent trials as a single aleatory experiment – a *compound experiment* that has $\mathcal{X}^n$ as its set of outcomes. The basic intuition underlying the doctrine of chances is that this compound experiment must be governed by the chance density $q^n : \mathcal{X}^n \to [0, 1]$ defined by

$$q^n(x_1, \cdots, x_n) = q(x_1) \cdots q(x_n) .$$

(Since q^n satisfies

$$\begin{aligned} \sum_{(x_1, \cdots, x_n) \in \mathcal{X}^n} q^n(x_1, \cdots, x_n) &= \sum_{(x_1, \cdots, x_n) \in \mathcal{X}^n} q(x_1) \cdots q(x_n) \\ &= \left(\sum_{x_1 \in \mathcal{X}} q(x_1) \right) \cdots \left(\sum_{x_n \in \mathcal{X}} q(x_n) \right) \\ &= 1 , \end{aligned}$$

it is indeed a chance density.) This chance density is called the product chance density for n independent trials governed by q.

The chance function associated with q^n may be denoted by $Ch^n : \mathcal{X}^n \to [0, 1]$; it assigns the chance

$$\mathrm{Ch}^n(U) = \sum_{(x_1,\cdots,x_n)\in U} q^n(x_1,\cdots,x_n) = \sum_{(x_1,\cdots,x_n)\in U} q(x_1)\cdots q(x_n)$$

to a subset U of $\mathfrak{X}^n$.

EXAMPLE 1.3. Since a single throw of one of Longcor's dice has the set of possible outcomes $\mathfrak{X} = \{1,2,3,4,5,6\}$, the compound experiment consisting of two physically independent throws has the set of outcomes $\mathfrak{X}^2$, which has 36 elements. If each single throw really is governed by the chance density q given in Example 1.2, then we would suppose a pair of physically independent throws to be governed by the corresponding product density q^2. If, for example, we were asked for the chance of a 6 followed by a 1, we would calculate the number

$$\begin{aligned} q^2(6,1) &= q(6)q(1) = (.179)(.155) \\ &= .028\,. \end{aligned}$$

And for the chance of the sum of the two throws being 3 or less, we would calculate

$$\begin{aligned} \mathrm{Ch}^2(\{(1,1),(1,2),(2,1)\}) &= q^2(1,1) + q^2(1,2) + q^2(2,1) \\ &= .073\,. \blacksquare \end{aligned}$$

The product chance density is fruitful because it permits one to incorporate into the mathematical theory itself ideas that begin only as intuitively understood features of randomness. The idea that the chance $q(x)$ is equal to the proportion of the time that x occurs becomes, for example, the *law of large numbers* – the theorem that as the number n of trials increases the chance approaches one that the proportion of the observed outcomes $x_1,\cdots,x_n$ equal to x will be approximated to given accuracy by $q(x)$.

The idea of a conditional chance law arises when an aleatory experiment has been partly completed and its outcome thus partly determined – partly determined in the sense that it will now necessarily fall in a subset U of the original set $\mathcal{X}$ of possible outcomes. In such a situation the role of chance seems not yet entirely played out; there is still a chance event involved in the determination of which element of U the outcome will be. But how does one calculate the chance that this residual chance event will produce a particular element x of U? Intuitively, this chance is equal to the proportion of the time that such a residual chance event does result in x – i.e., to the proportion of those outcomes falling in U that finally fall equal to x. And this proportion is given by the ratio

$$\frac{q(x)}{Ch(U)} = \frac{Ch(\{x\})}{Ch(U)}$$

where q is the original chance density and Ch is its chance function.

So "conditioning" on a subset U of the set of possibilities of a chance density $q : \mathcal{X} \to [0,1]$ leads to a *conditional density*, the density $q_U : \mathcal{X} \to [0,1]$ defined by

$$q_U(x) = \begin{cases} \dfrac{q(x)}{Ch(U)} & \text{if} \quad x \in U \\ 0 & \text{if} \quad x \notin U , \end{cases}$$

where Ch is the chance function for q. Of course, we can "condition on U" only if $Ch(U) > 0$ – i.e., q must permit U to occur. (In this case,

$$\sum_{x \in \mathcal{X}} q_U(x) = \sum_{x \in U} \frac{q(x)}{Ch(U)} = \frac{Ch(U)}{Ch(U)} = 1 ,$$

so that q_U is indeed a chance density.)

The chance function associated with the conditional chance density q_U may be denoted Ch_U; it is given by

$$Ch_U(V) = \sum_{x \in V} q_U(x) = \sum_{x \in U \cap V} \frac{q(x)}{Ch(U)} = \frac{Ch(U \cap V)}{Ch(U)}$$

for all $V \subset \mathcal{X}$. The chance $Ch_U(V)$, which may also be written $Ch(V|U)$, is called the *conditional chance* of V given U. And the formula

$$Ch(V|U) = \frac{Ch(U \cap V)}{Ch(U)}$$

is called the *rule of conditioning* for chances.

EXAMPLE 1.4. *Inverse Sampling.* A fair coin is to be tossed until two heads appear. The set $\mathcal{X}$ of possible outcomes of this experiment is the set of all sequences of heads and tails that include exactly two heads, one of them at the end. For example: HTTH is in $\mathcal{X}$, but HHT and THTHH are not. Since the coin is "fair," the chance density $q: \mathcal{X} \to [0,1]$ assigns each element $x \in \mathcal{X}$ the chance 2^{-n}, where n is the length of the sequence x. For example: $q(HTTH) = \frac{1}{16}$.

We begin the experiment: we toss the coin once and a head appears. The situation has changed; no sequence in $\mathcal{X}$ that begins with T is now possible, and one that begins with H now has chance $2^{-(n-1)}$, where n is its length. For example, the chance of HTTH is $\frac{1}{8}$.

This transformation may be described, of course, by the process of conditioning. The chance density must be conditioned on the set U consisting of all sequences beginning with H, and since $Ch(U) = \frac{1}{2}$, this conditioning yields

$$q_U(x) = \begin{cases} 2q(x) & \text{if } x \in U \\ 0 & \text{if } x \notin U \end{cases}$$

$$= \begin{cases} 2^{-(n-1)} & \text{if } x \text{ begins with H, where n is the length of } x \\ 0 & \text{if } x \text{ does not begin with H.} \end{cases} \blacksquare$$

Conditional chance has always played a relatively minor role in the mathematical theory of chance. Its importance has increased in recent decades, especially in the part of the theory concerned with "stochastic processes"; but its role is still small relative to the role played by the idea of product densities. As we will see shortly, "the rule of conditioning" figures most prominently not in the theory of chance but in a theory of partial belief.

§4. Chances as Degrees of Belief

The chances governing an aleatory experiment may or may not coincide with our degrees of belief about the outcome of the experiment. If we know the chances, then we will surely adopt them as our degrees of belief. But if we do not know the chances, then it will be an extraordinary coincidence for our degrees of belief to be equal to them.

The second case is the typical one. When we first conceive of an experiment as random, we typically have little idea about what chance density governs it. And though we may eventually form some opinion about the true chance density on the basis of actual observations of the experiment, we may never obtain any very exact or certain values for the true chances. Even after an immense number of observations, we may only have, as in the case of Longcor's dime-store dice, a guess about the true chances based on speculative assumptions.

Furthermore, scientific applications of the theory of chance usually turn on the fact that the chance density governing an experiment is, in the first instance, unknown. Typically, a scientist is interested in an aleatory experiment precisely because it might be governed by any one of several chance densities, each of which is associated with some hypothesis of scientific interest. In such a case observation of the experiment may provide evidence as to which chance density – and hence which hypothesis – is correct.

Chances, then, must be conceived of as features of the world. They are not necessarily features of our knowledge or belief. And it would be quite untenable to claim that a chance is *merely* a feature of our knowledge or belief. Yet such a claim has often been made.

The most influential advocate of such a claim was probably Pierre Simon Laplace (1749-1827), the French mathematician famous both for his development of the mathematical theory of chance and for his insistence on determinism in science. In his practice of statistical inference, Laplace always carefully distinguished between chances and the degrees of belief that he sometimes combined with those chances according to the Bayesian method. (See §10 below.) But as a determinist he could not make philosophical sense of randomness, and he insisted, therefore, that chance is merely a feature of our knowledge, papering over as best he could the conflict between this doctrine and the fact that chances may be unknown. Since the advent of quantum mechanics, Laplace's determinism has lost its grip on physics, but his attitude toward chance has not yet entirely disappeared.

A tendency to reduce chances to degrees of belief can also arise within the mathematical theory of chance itself. For many of the ideas of that theory can be exposited in terms of partial belief, and some of them are most easily exposited in this way. Consider, for example, the method of conditioning. In the previous section this method appeared as a method for describing the partial completion of an aleatory experiment: such a partial completion determines that the outcome will be in a set U and thus results in the replacement of the original chance density q by a new chance density that can be calculated by conditioning on U. But the purely mathematical potential of the method outstrips this aleatory picture. For the method can be formally applied to almost any subset of an aleatory experiment's set of possible outcomes; yet we will usually find few ways that such an experiment can be partially completed and hence few subsets for which conditioning will make aleatory sense. Conditioning on other subsets will make sense only in terms of degrees of belief.

EXAMPLE 1.5. *Epistemic Conditioning*. Consider the toss of a balanced die, with its set of possible outcomes $\mathcal{X} = \{1,2,3,4,5,6\}$ and its chance density $q : \mathcal{X} \to [0,1]$ given by $q(x) = \frac{1}{6}$ for all $x \in \mathcal{X}$. There seems no way that such a toss can be only partially completed.

It can make sense, though, to condition q on a subset of $\mathcal{X}$. For suppose we know q to govern the toss and we adopt the chances given by q as our degrees of belief about what the outcome will be. And suppose that the toss is made and we receive only a partial report on how it came out – we are told, say, that it came out even. Then it is natural to modify our degrees of belief by conditioning q on the subset $U = \{2,4,6\}$ of even outcomes. In other words, we will adopt the chances

$$q_U(x) = \begin{cases} \frac{1}{3} & \text{for} \quad x = 2,4,6 \\ 0 & \text{for} \quad x = 1,3,5 \end{cases}$$

as our degrees of belief about how the toss actually came out. ■

Fortunately, the opinion that chances are merely degrees of belief is no longer widely held. But the historical prevalence of this doctrine and the ease with which chances can be interpreted as degrees of belief have both contributed to another idea that I believe to be equally in error: the idea that degrees of belief ought always to be like chances in their mathematical structure. It is to this idea that we now turn.

§5. The Bayesian Theory of Partial Belief

As I have stressed in the preceding pages, the degrees of belief studied in this essay do not, in general, obey all the rules for chances. But there is an extensively developed and very popular theory of partial belief, usually called the *Bayesian theory* after the English clergyman Thomas Bayes (1702-1761), that begins with the explicit premise that all degrees of belief should obey these rules. The Bayesian theory adopts the three basic rules for chances as rules for one's degrees of belief based on a given body of evidence, and it adopts the rule of conditioning for chances as a general rule for changing one's degrees of belief when that evidence is augmented by the knowledge of a particular proposition.

The first three rules of the Bayesian theory can be expressed as rules governing a set function: one's degrees of belief with respect to a set of possibilities Θ must be given by a function $\text{Bel}: 2^{\Theta} \to [0,1]$ that obeys three rules:

(1) $\text{Bel}(\emptyset) = 0$.

(2) $\text{Bel}(\Theta) = 1$.

(3) If $A \cap B = \emptyset$, then $\text{Bel}(A \cup B) = \text{Bel}(A) + \text{Bel}(B)$.

Notice that rules (1) and (2) are the same as the first two rules for belief functions, while rule (3) differs from the third rule for belief functions. Rule (3) is called *Bayes' rule of additivity.*

As we will see in §6 of Chapter 2, a function $\text{Bel}: 2^{\Theta} \to [0,1]$ that obeys the three preceding rules necessarily obeys the three rules for belief functions – i.e., it is a belief function. But not all belief functions obey Bayes' rule of additivity. Thus the functions that obey the three Bayesian rules form a proper subclass of the class of belief functions. I call them *Bayesian belief functions.*

The fourth rule of the Bayesian theory says that if we begin with a Bayesian belief function $\text{Bel}: 2^{\Theta} \to [0,1]$ and then learn that $A \subset \Theta$ is true, then we should replace Bel with a new Bayesian belief function which may be denoted by $\text{Bel}_A : 2^{\Theta} \to [0,1]$ and which is given by

$$\text{Bel}_A(B) = \frac{\text{Bel}(B \cap A)}{\text{Bel}(A)} \tag{1.3}$$

for all $B \subset \Theta$. (Here it is assumed that $\text{Bel}(A) > 0$. Under this assumption, (1.3) does indeed define a Bayesian belief function.) In most Bayesian accounts, the quantity $\text{Bel}_A(B)$ is denoted by $\text{Bel}(B|A)$, so that the rule becomes

(4) If $\mathrm{Bel}(A) > 0$, then $\mathrm{Bel}(B|A) = \dfrac{\mathrm{Bel}(B \cap A)}{\mathrm{Bel}(A)}$.

The expression $\mathrm{Bel}(B|A)$ is read "the conditional degree of belief in B given A," and (4) is called *Bayes' rule of conditioning.*

Just as the Bayesian belief functions are a subclass of our belief functions, so Bayes' rule of conditioning is a special case of Dempster's rule of combination. This is made clear in §5 of Chapter 3 below, where it is shown that Bel_A is equal to the orthogonal sum of the Bayesian belief function Bel with the simple support function that focuses unit support on the subset A.

The Bayesian theory is thus contained in our theory as a restrictive special case. In the next several sections of this chapter, I outline my reasons for thinking that it is too restrictive a special case – that the greater flexibility of our theory is valuable and even essential for an adequate representation of evidence and probable reasoning.

§6. The Role of Judgment

Whenever I write in this essay of the "degree of support" that given evidence provides for a proposition or of the "degree of belief" that an individual accords the proposition, I picture in my mind an act of judgment. I do not pretend that there exists an objective relation between given evidence and a given proposition that determines a precise numerical degree of support. Nor do I pretend that an actual human being's state of mind with respect to a proposition can ever be described by a precise real number called his degree of belief, nor even that it can ever determine such a number. Rather, I merely suppose that an individual can make a judgment. Having surveyed the sometimes vague and sometimes confused perception and understanding that constitutes a given body of evidence, he can announce a number that represents the degree to which he judges that evidence to support a given proposition and, hence, the degree of belief he wishes to accord the proposition.

Unfortunately, this view of the proper roles of evidence and judgment in determining degrees of support and belief is shared by few contemporary proponents of the Bayesian theory. In fact, almost all twentieth century Bayesians have explicitly advocated one or the other of the two alternatives I have just rejected. Either they have followed Harold Jeffreys and John Maynard Keynes in insisting that numerical degrees of support are indeed objectively determined by given evidence, or else they have followed Frank Plumpton Ramsey and Bruno de Finetti in choosing to analyze degrees of belief as psychological facts, facts which can be discovered by observing an individual's preferences among bets or risks but which may not bear any particular relation to any particular evidence.

These two Bayesian views are usually labelled the *logical view* and the *personalist view*, respectively. The logical view is the older of the two, and the one that is most akin to eighteenth and nineteenth century Bayesian thinking. But since the Second World War, and especially since L. J. Savage published his *Foundations of Statistics* in 1954, personalism has become the predominant Bayesian opinion – so much so that the adjectives "personalist" and "Bayesian" are now often used, with some justice, as synonyms.

In the next two sections we will gain some understanding of why Bayesians have felt compelled to adopt either the logical view or the personalist view, and of why the personalist view has proven the more viable one. But the prominence of the personalist view requires that we first pause to remark further on the divergence of that view from our present perspective.

As I have already remarked, personalists do not seek to analyze the relation between an individual's degrees of belief and his evidence. Nor do they seek to relate the structure of those degrees of belief to the nature of evidence. Instead they set themselves the task of finding conditions that a set of degrees of belief must obey in order to be internally consistent. And they advance various arguments to the effect that the Bayesian rules provide just such conditions – that an individual will be "irrational," "incoherent," or "inconsistent" (the epithet varies) if his degrees of belief fail to obey the Bayesian rules.

Because of its de-emphasis of the notion of evidence, the personalist view fits awkwardly into the present exposition. It is not clear, for example, whether it is fair to the personalist view to say that the first three Bayesian rules apply to degrees of belief "based on a given body of evidence." I hasten, therefore, to say that my description of the Bayesian theory in the preceding section is accurate only to the extent that the theory is construed as a theory of evidence.

I reject without apology, however, the personalist contention that non-Bayesian degrees of belief such as those studied in this essay are incoherent. Though this is not the place to exposit and refute the various personalist arguments on this topic, I must say that I consider those arguments unsound. Indeed, I think it obvious on the face of things that such arguments cannot possibly achieve their goal.

§7. The Representation of Ignorance

As we have already seen in the example of the Ming vase, belief functions readily lend themselves to the representation of ignorance. The rules for belief functions permit us, when we have little evidence bearing on a proposition, to express frank agnosticism by according both that proposition and its negation very low degrees of belief. As a matter of fact, these rules permit us to profess a zero degree of belief for every proper subset of a set of possibilities Θ. That is to say, the function $\mathrm{Bel}: 2^{\Theta} \to [0,1]$ defined by

$$\mathrm{Bel}(A) = \begin{cases} 0 & \text{if} \quad A \neq \Theta \\ 1 & \text{if} \quad A = \Theta \end{cases}$$

qualifies as a belief function. It is called the *vacuous belief function*, and it is obviously the belief function we would use to represent complete ignorance – i.e., to represent the situation where we have no evidence about Θ at all.

The Bayesian theory, on the other hand, cannot deal so readily with the representation of ignorance, and it has often been criticized on this account. The basic difficulty is that the theory cannot distinguish between lack of belief and disbelief. It does not allow one to withhold belief from a proposition without according that belief to the negation of the proposition.

Indeed, whenever a proposition is represented by a subset A of Θ, its negation is represented by the complement $\bar{A}$ of A – i.e., by the subset of Θ consisting of all elements not in A. And since $A \cup \bar{A} = \Theta$, rule (2) of the Bayesian theory requires that $Bel(A \cup \bar{A}) = 1$. And Bayes' rule of additivity, therefore, requires that

$$Bel(A) + Bel(\bar{A}) = 1 . \tag{1.4}$$

This implies that Bel(A) cannot be low unless $Bel(\bar{A})$ is high: failure to believe A necessitates accordance of belief to $\bar{A}$.

Now (1.4) does permit one to set

$$Bel(A) = Bel(\bar{A}) = \frac{1}{2} ,$$

and it is sometimes argued that doing so is a way of expressing complete ignorance about A within the Bayesian theory. Ignorance, it is argued, is represented by a degree of belief equal to $\frac{1}{2}$. But though such a device seems plausible in the case of a single dichotomy $(A, \bar{A})$, it is useless in representing ignorance with respect to a set of possibilities Θ containing more than two elements. For the Bayesian theory does not allow one to give degree of belief $\frac{1}{2}$ to every proper subset of such a set.

EXAMPLE 1.6. *Life Near Sirius?* Are there or are there not living beings in orbit around the star Sirius? Some scientists may have evidence on this question, but most of us will profess complete ignorance about it. So if θ_1 denotes the possibility that there is such life and θ_2 denotes the possibility that there is not, we will adopt the vacuous belief function over the set of possibilities $\Theta = \{\theta_1, \theta_2\}$.

We can also consider the question in the context of a more refined set of possibilities. We might, for example, raise the question of whether there even exist planets around Sirius. We would then have a set of possibilities $\Omega = \{\zeta_1, \zeta_2, \zeta_3\}$, say, where ζ_1 corresponds to the possibility that there is life around Sirius, ζ_2 corresponds to the possibility that there are planets but no life, and ζ_3 corresponds to the possibility that there are not even planets. (The set Ω is related to the set Θ in that ζ_1 corresponds to θ_1 and $\{\zeta_2, \zeta_3\}$ corresponds to θ_2.) Many of us will adopt the vacuous belief function over Ω, and this will be consistent with having adopted the vacuous belief function over Θ.

The Bayesian will find it difficult to specify consistent degrees of belief over Θ and Ω that he can defend as a representation of ignorance. Focusing on Θ, he might claim that ignorance is represented by

$$\mathrm{Bel}(\{\theta_1\}) = \mathrm{Bel}(\{\theta_2\}) = \frac{1}{2} . \tag{1.5}$$

But when he turns to Ω, he has to satisfy

$$\mathrm{Bel}(\{\zeta_1\}) + \mathrm{Bel}(\{\zeta_2\}) + \mathrm{Bel}(\{\zeta_3\}) = 1 ,$$

and thus the best he can do to represent ignorance among the three alternatives is to set

$$\mathrm{Bel}(\{\zeta_1\}) = \mathrm{Bel}(\{\zeta_2\}) = \mathrm{Bel}(\{\zeta_3\}) = \frac{1}{3} .$$

But this yields

$$\mathrm{Bel}(\{\zeta_1\}) = \frac{1}{3} , \quad \mathrm{Bel}(\{\zeta_2, \zeta_3\}) = \frac{2}{3} . \tag{1.6}$$

And since $\{\theta_1\}$ has the same meaning as $\{\zeta_1\}$ and $\{\theta_2\}$ has the same meaning as $\{\zeta_2, \zeta_3\}$, (1.5) and (1.6) are inconsistent. ■

It should be stressed that there is no novelty in criticizing the Bayesian theory for its inability to represent ignorance. Such criticism was an important factor in the decline of Bayesian ideas in the nineteenth century,* and it figures today in textbook discussions of the Bayesian theory. Contemporary proponents of the personalist version of the Bayesian theory, secure in their conviction that the degrees of belief of rational individuals must obey the Bayesian rules, conclude only that it is irrational to profess complete ignorance.

§8. Combination vs. Conditioning

Dempster's rule of combination provides a method for changing prior opinions in the light of new evidence: we construct a belief function to represent the new evidence and combine it with our "prior" belief function – i.e., with the belief function that represents our prior opinions. This method deals symmetrically with the new evidence and the old evidence on which our prior opinions are based: both bodies of evidence are represented by belief functions, and the result of the combination does not depend on which evidence is the old and which is the new.

In the Bayesian theory, the task of telling how our degrees of belief ought to change as new evidence is obtained falls to Bayes' rule of conditioning: we represent the new evidence as a proposition and condition our prior Bayesian belief function on that proposition. Here we find no obvious symmetry in the treatment of new and old evidence. And more importantly, we find that the assimilation of new evidence depends on an astonishing assumption: we must assume that the exact and full effect of that new evidence is to establish a single proposition with certainty. In

*The story of the decline of the Bayesian doctrine after the death of Laplace is a complicated one, and it has never been adequately told. But certainly one of the most forceful of the early critics of the doctrine was George Boole, who stressed the arbitrary character of "*a priori* probabilities." The quotation at the head of Chapter 11 is taken from page 375 of Boole's *Laws of Thought*, published in 1854.

A brief account of the decline of the Bayesian doctrine in England can be found in Chapter II of R. A. Fisher's *Statistical Methods and Scientific Inference*.

contrast to Dempster's rule of combination, which can accommodate new evidence that justifies only partial beliefs, Bayes' rule of conditioning requires that the new evidence be expressible as a certainty.*

Few Bayesians have ever questioned the assumption that evidence can always be expressed as a certainty.** The assumption is, after all, a philosophical commonplace – one that has been accepted by generations of epistemologists. Nevertheless, it is the source of a crucial difficulty for the Bayesian theory. In fact, it can be held responsible for the bifurcation of the Bayesian school of thought into its unviable "logical" branch and its anti-evidential "personalist" branch.

The difficulty lies in the need to reconcile the assumption that evidence can always be expressed as a certainty with the fact that the prior Bayesian belief function, itself presumably based on evidence, usually expresses not only certainties but also positive degrees of belief falling short of certainty. There appears to be only two ways of effecting such a reconciliation. Either one must accept the idea that the prior Bayesian belief function is based on evidence, as the proponents of the logical view do, and then argue that the proposition that expresses that evidence can provide positive but partial support for other propositions. Or else one must follow the personalists in retreating from the idea that the prior Bayesian belief function is based on any particular evidence.

The argument advanced by the proponents of the logical view may at first glance seem tenable. But it has not proven popular among contemporary

*Since Bayes' rule of conditioning can be applied repeatedly, it can also deal with new evidence whose exact and full effect is to establish a collection of several propositions. But a collection of propositions is always equivalent to a single proposition – their conjunction.

**One of the few is Richard Jeffrey, who has advanced a generalization of Bayes' rule that takes into account the fact that the new evidence may justify only partial beliefs. Unfortunately, this generalization still treats the old and new evidence asymmetrically, and hence cannot qualify as a genuine rule of combination. See Chapter 11 of Jeffrey's *Logic of Decision*.

students of the Bayesian theory. Most of these students, myself included, find it difficult to see how so slight and formal an object as a proposition can, in itself and without reference to the experience by which we come to know it, provide positive but partial support to other propositions. This is true whether one thinks of a proposition as a sentence in a formal language, as logicians usually do, or as a subset of possibilities, as I do in this essay. When we think of two propositions as sentences in a formal language, we find no relation between them save by virtue of their grammar, and we are reduced to the unworkable proposal that such grammatical relations are the source of degrees of support. (Rudolf Carnap, the most persistent recent proponent of the logical view, made precisely this proposal in his *Continuum of Inductive Methods*, published in 1952.) And when we think of a proposition as a subset of possibilities, we find that the proposition leaves us totally ignorant about which element of that subset is correct, and we face the insoluble problem of finding Bayesian degrees of support to express this total ignorance.

By liberating the prior Bayesian belief function from the notion of evidence, the personalist view escapes the difficulty that defeats the logical view. But in asking us to provide a non-evidential prior Bayesian belief function that can be conditioned on the proposition representing our new evidence, it often presents us with a formidable task. For Bayes' rule of conditioning not only requires that the new evidence be represented as a proposition – it also requires that that proposition correspond to a subset of the set of possibilities Θ over which the prior Bayesian belief function is defined. And an attempt to express our evidence as a proposition will often lead to a proposition that is grotesquely complicated and hence necessitate the use of a set Θ that is so immense and detailed as to outstrip any prior opinions that we might actually have had.

Since it does not require us to express our evidence as a certainty, Dempster's rule of combination permits us to construct descriptions of probable reasoning that are more modest than such Bayesian descriptions but more faithful to the way human beings actually think. For instead of forcing us to pretend to detailed prior opinions over a set of possibilities whose very formulation reaches or exceeds the limits of our understanding, it permits us to approach a problem by formulating a small, well-understood set of possibilities and then using the balance of our understanding to assess appropriate prior degrees of belief over that set.

§9. The Representation of Probable Reasoning

When we combine evidence supporting one proposition with evidence supporting a compatible proposition, we obtain support for the conjunction of the two propositions. If, for example, we have testimony that a thief was dark-haired and have other evidence pointing to the thief's being left-handed, then we may conclude that there is some support for the thief's being a dark-haired left-hander. This most elementary type of probable reasoning can be represented quite naturally in our theory.

Suppose, indeed, that A and B are two subsets of a set of possibilities Θ, and suppose that A and B are compatible – i.e., $A \cap B \neq \emptyset$. And suppose we have two items of evidence, one pointing precisely to A and one pointing precisely to B. Then we will, of course, represent the first item of evidence by a simple support function focused on A, represent the second body of evidence by a simple support function focused on B, and combine the two by Dempster's rule. As we will see in §4 of Chapter 4, the resulting separable support function does provide support for the conjunction $A \cap B$; if the first simple support function awards its focus A the degree of support s_1 and the second simple support function awards its focus B the degree of support s_2, then the orthogonal sum will award the conjunction $A \cap B$ the degree of support $s_1 s_2$.

The representation of probable reasoning is more problematic in the Bayesian theory. This is partly because devices like simple support functions are unavailable, but more fundamentally because there is no way to isolate the opinions one wants to combine from the prior Bayesian belief function and the mass of other opinions it is supposed to represent. The "prior degrees of belief" inevitably dominate any calculation.

EXAMPLE 1.7. Consider the problem of combining evidence that supports a proposition A with evidence that conclusively establishes a compatible proposition B. The representation in terms of simple support functions will show all the original support for A devolving on the conjunction $A \cap B$. But in a Bayesian representation such a result will be contingent on the details of the Bayesian prior.

Begin, indeed, with an item of evidence that points to a subset A of a set of possibilities Θ, and contemplate combining it with possible future evidence representable as some compatible subset of Θ. According to the Bayesian theory, you will incorporate the evidence pointing to A in a prior Bayesian belief function $\mathrm{Bel}: 2^{\Theta} \to [0,1]$ which awards a high degree of belief to A. But the high value of Bel(A) will hardly guarantee that conditioning on a compatible subset B will produce a high degree of belief for $A \cap B$. In fact, if A has sufficiently many elements relative to the size of Bel(A), then B can be chosen so as to make the conditional degree of belief for $A \cap B$ arbitrarily *small*!

(Suppose the number of elements of A is many times larger than $(\mathrm{Bel}(\bar{A}))^{-1}$. Then an element $\theta \in A$ may be chosen so that $\mathrm{Bel}(\{\theta\})$ is many times smaller than $\mathrm{Bel}(\bar{A})$. And the choice $B = \{\theta\} \cup \bar{A}$ yields

$$\mathrm{Bel}(A \cap B|B) = \frac{\mathrm{Bel}(A \cap B)}{\mathrm{Bel}(B)} = \frac{\mathrm{Bel}(\{\theta\})}{\mathrm{Bel}(\{\theta\}) + \mathrm{Bel}(\bar{A})}$$

$$\leq \frac{\mathrm{Bel}(\{\theta\})}{\mathrm{Bel}(\bar{A})},$$

and this is very small.) ■

§10. Statistical Inference

Given one or more observations from a random experiment, what can we say about the chance density that governs the experiment? This problem, the problem of *statistical inference*, has always been as central to the development of theories of partial belief as it has been to scientific applications of the notion of chance. In fact, a solution to the problem of statistical inference was the object both of Thomas Bayes' essay and of Arthur Dempster's work on upper and lower probabilities. The present essay, though it does not claim to offer a thorough or definitive treatment of the problem, does advance some proposals that may merit comparison with the proposals advanced by the Bayesian and other theories.

It is easy to see why statistical inference holds such interest for a theory of partial belief. Suppose $\mathcal{X}$ is the set of possible outcomes for some random experiment, and suppose we know the experiment is governed by one of a class $\{q_\theta\}_{\theta\in\Theta}$ of chance densities on $\mathcal{X}$. Then it would seem that an observed outcome $x\in\mathcal{X}$, since it is evidence as to which q_θ governs the process, ought to determine a belief function S_x over Θ. And the wealth of precise mathematical information contained in the chance densities $\{q_\theta\}_{\theta\in\Theta}$ permits one to hope that such a belief function can in fact be determined by some general convention rather than by intuitive judgments directed at the particular case.

The problem I have just described may be called the problem of *statistical estimation*. The approach to it that is most natural from the perspective of our theory is to formulate a convention as to what weights of evidence ought to be associated with an observation x. Several such conventions are possible; one is advanced and briefly explored in Chapter 11.

The Bayesian approach to the problem of statistical estimation is well-known. It begins, as always, with a prior Bayesian belief function $Bel_0 : 2^\Theta \rightarrow [0,1]$. But since it requires a set of possibilities in which the evidence can appear as a subset, it temporarily abandons Θ in favor of the Cartesian product $\Theta \times \mathcal{X}$, an element (θ, x) of which corresponds to the possibility, before the experiment is conducted, that q_θ is the true chance density *and* x will be the outcome. The proposition that x is observed does correspond to a subset of $\Theta \times \mathcal{X}$ – namely, to the subset

$\Theta \times \{x\}$. Moreover, the prior Bayesian belief function $\mathrm{Bel}_0 : 2^\Theta \to [0,1]$ extends in a natural way to a Bayesian belief function over $\Theta \times \mathcal{X}$. This is because there is only one Bayesian belief function $\mathrm{Bel} : 2^{\Theta\times\mathcal{X}} \to [0,1]$ that satisfies the quite natural conditions that

(1) $\mathrm{Bel}(A \times \mathcal{X}) = \mathrm{Bel}_0(A)$ for all $A \subset \Theta$,

and

(2) $\mathrm{Bel}(\Theta \times \{x\} | \{\theta\} \times \mathcal{X}) = q_\theta(x)$ for all $\theta \in \Theta$, $x \in \mathcal{X}$.

(Since $A \times \mathcal{X}$ corresponds to the same proposition as A, (1) is unexceptionable. And since $\{\theta\} \times \mathcal{X}$ corresponds to the proposition that q_θ is the true chance density, (2) is merely the Bayesian way of expressing the truism that when we know a chance law we should adopt the chances it gives as our degrees of belief.) After the Bayesian observes x, he will of course condition Bel on $\Theta \times \{x\}$, thus obtaining new degrees of belief over $\Theta \times \mathcal{X}$. And he will thus obtain a "posterior" Bayesian belief function over Θ, which may be denoted by $\mathrm{Bel}_1 : 2^\Theta \to [0,1]$, and which will be given by

$$\mathrm{Bel}_1(A) = \mathrm{Bel}(A \times \mathcal{X} | \Theta \times \{x\})$$

for all $A \subset \Theta$.

By Bayes' rule of conditioning,

$$\mathrm{Bel}(\Theta \times \{x\} | \{\theta\} \times \mathcal{X}) = \frac{\mathrm{Bel}(\{\theta\} \times \{x\})}{\mathrm{Bel}(\{\theta\} \times \mathcal{X})}$$

and

$$\mathrm{Bel}(\{\theta\} \times \mathcal{X} | \Theta \times \{x\}) = \frac{\mathrm{Bel}(\{\theta\} \times \{x\})}{\mathrm{Bel}(\Theta \times \{x\})}.$$

Hence the posterior Bayesian belief function Bel_1 assigns any particular $\theta \in \Theta$ the degree of belief

$$\begin{aligned}
\mathrm{Bel}_1(\{\theta\}) &= \mathrm{Bel}(\{\theta\} \times \mathcal{X} | \Theta \times \{x\}) \\
&= \frac{\mathrm{Bel}(\{\theta\} \times \{x\})}{\mathrm{Bel}(\Theta \times \{x\})} = \frac{\mathrm{Bel}(\{\theta\} \times \mathcal{X})\,\mathrm{Bel}(\Theta \times \{x\} | \{\theta\} \times \mathcal{X})}{\mathrm{Bel}(\Theta \times \{x\})} \\
&= \frac{\mathrm{Bel}_0(\{\theta\})\, q_\theta(x)}{\mathrm{Bel}(\Theta \times \{x\})} \\
&= K\, \mathrm{Bel}_0(\{\theta\})\, q_\theta(x),
\end{aligned}$$

where the constant K does not depend on θ. The formula

$$\mathrm{Bel}_1(\{\theta\}) = K\, \mathrm{Bel}_0(\{\theta\})\, q_\theta(x) \qquad (1.7)$$

is often called *Bayes' Theorem.*

The Bayesian solution to the problem of statistical estimation has much to commend it, and it has always been the main source of the appeal of the Bayesian theory as a whole. Indeed, statisticians have almost unanimously agreed that it is the right solution *for someone who actually has a Bayesian prior*, and most have refused to accept it as a general solution to the problem only because they have felt that such a prior is usually lacking.

As we will see in Chapter 11, our method for determining a belief function S_x to represent the effect of the observation x does not conflict with the Bayesian solution to the problem. For when the belief function S_x is combined with a Bayesian prior by Dempster's rule of combination, it produces the corresponding Bayesian posterior. But it is not necessary to introduce such a prior. S_x may be combined instead with a non-Bayesian prior belief function representing our prior opinions, or else it may be left to stand alone as an expression of the degrees of support actually provided by the observation.

§11. The Bayesian Theory as a Limiting Case

As we have seen, the Bayesian theory is a special case of the theory of this essay. The Bayesian belief functions are a subset of our belief functions, Bayes' rule of conditioning is a special case of Dempster's rule of combination, and the Bayesian solution to the problem of statistical estimation can be obtained by using Dempster's rule to combine the Bayesian prior with a belief function representing the statistical evidence. Only one further remark is needed to complete the picture: Bayesian belief functions are usually quasi support functions.

As we learn in §3 of Chapter 9, the only Bayesian belief functions that qualify as support functions are those that are trivially Bayesian –

those that concentrate unit belief on a single element of the set of possibilities. The non-trivially Bayesian belief functions are all quasi support functions. The Bayesian theory emerges, therefore, as a limiting case of our theory – a limiting case in which fictional infinite contradictory weights of evidence suppress much of the detail of one's actual evidence.

§12. Probability

In the preceding pages I have avoided the word "probability." It is a beautiful word, rich in history and in connotation. But it is an ambiguous word; it is widely used both for the *aleatory* concept that I have been calling "chance" and for the *epistemic* concept that I have been calling "degree of belief" or "degree of support." This ambiguity has, in my opinion, been the source of much of the support for the Bayesian theory. And it has undeniably been the source of endless confusion in the teaching of statistical inference.

Careful writers have managed, of course, to distinguish between the two meanings of "probability." They distinguish between *objective* and *subjective* probabilities, or else between *aleatory* and *epistemic* probabilities. But these clumsy adjectives usually fail to persist in lengthy discussions, and they seldom even appear in the elementary discussions to which students of statistics are first exposed.

The epistemic sense of "probability" is, of course, the older one. Before the rise of the mathematical theory of chances in the late seventeenth century, no European would have thought to connect judgments about whether an opinion is "probable" with the idea of blind chance. But once the connection was made – once the chance of an event in a game of chance was recognized as a measure of the probability of the opinion that the event will occur – it was not long before "probability" acquired its aleatory connotations.*

*For an extended discussion of the development of the word "probability" in the seventeenth century, see Ian Hacking's *Emergence of Probability*.

How should we use the word "probability" today? Should we retain the well-established mathematical and scientific practice of using the word to mean chance? Or should we bow to the ancient, etymologically correct, and still living epistemic usage? Most likely we will continue to compromise; we will use the word to mean anything and everything in most of our teaching and conversation, and we will distinguish between aleatory probability and epistemic probability only in our more careful and advanced discussions. My own preference, though, is that we should finally adopt the convention proposed by Poisson: use *chance* for the aleatory concept, and reserve *probability* for the epistemic concept. Such a convention would spare every student of statistical inference a good deal of needless confusion.

CHAPTER 2. DEGREES OF BELIEF

> *Probabilitas* enim est gradus certitudinis, & ab hac differt ut pars à toto.
>
> JAMES BERNOULLI (1654-1705)

The additive degrees of belief of the Bayesian theory correspond to an intuitive picture in which one's total belief is susceptible of division into various portions, and that intuitive picture has two fundamental features. First, to have a degree of belief in a proposition is to commit a portion of one's belief to it. And secondly, whenever one commits only a portion of one's belief to a proposition, one must commit the remainder to its negation.

The obvious way to obtain a more flexible and realistic picture is to discard the second of these features while retaining the first. As we see in this chapter, this leads to the theory of *belief functions*, a mathematical theory that is flexible yet richly structured.*

§1. Subsets as Propositions

Before introducing the belief functions, let us review the formalism whereby propositions are represented as subsets of a given set.

---

*The set functions here called belief functions were studied by Arthur Dempster in his 1967 paper "Upper and lower probabilities induced by a multivalued mapping." They had earlier been studied extensively, but without reference to the notion of partial belief, by the French mathematician Gustave Choquet, who called them "monotone of order infinity," and by his student André Revuz, who called them "totalement croissantes"; see their articles in *Annales de l'Institut Fourier*. Since Dempster's work, other statisticians have studied related classes of functions from somewhat different points of view; see the articles by Beran, Huber, and Huber and Strassen.

This formalism is most easily introduced in the case where we are concerned with the true value of some quantity. If we denote the quantity by θ and the set of its possible values by Θ, then the propositions of interest are precisely those of the form "The true value of θ is in T," where T is a subset of Θ. Thus the propositions of interest are in a one-to-one correspondence with the subsets of Θ, and the set of all propositions of interest corresponds to the set of all subsets of Θ, which is denoted by the symbol 2^{Θ}.

The same formalism can also be extended to situations where we are not explicitly concerned with such a quantity θ. For if we allow θ to be an arbitrary "parameter" that takes possibly non-numerical values, then θ and Θ can be chosen so that 2^{Θ} will include any particular fixed set of propositions we wish it to. If, for example, we wish 2^{Θ} to include all propositions delimiting the date and/or place of origin of a particular relic, then we need only let θ be the "date and place of origin of the relic," and let Θ be the set of all pairs consisting of a possible date and a possible place of origin. And we can similarly deal with almost any example of probable reasoning by letting Θ be the set of all the different possibilities under consideration.

It should not be thought that the "possibilities" that comprise Θ will be determined and meaningful independently of our knowledge. Quite to the contrary: Θ will acquire its meaning from what we know or think we know; the distinctions that it embodies will be embedded within the matrix of our language and its associated conceptual structures and will depend on those structures for whatever accuracy and meaningfulness they possess. In order to emphasize this epistemic nature of the set of possibilities Θ, I will call it the *frame of discernment*. When a proposition corresponds to a subset of a frame of discernment, I will say that the frame *discerns* that proposition.

One reason that the correspondence between propositions and subsets is useful is that it translates the logical notions of conjunction, disjunction, implication and negation into the more graphic set-theoretic notions of intersection, union, inclusion and complementation. Indeed, if A and B are two subsets of Θ, and A′ and B′ are the corresponding propositions, then the intersection $A \cap B$ corresponds to the conjunction of A′ and B′, the union $A \cup B$ corresponds to the disjunction of A′ and B′, A is a subset of B (written $A \subset B$) if and only if A′ implies B′, and A is the set-theoretic complement of B with respect to Θ (written $A = \bar{B}$) if and only if A′ is the negation of B′. Notice also that 2^{Θ} includes the empty set $\emptyset$, which corresponds to a proposition that is known to be false, and the whole set Θ, which corresponds to a proposition that is known to be true.

For the sake of mathematical simplicity, the discussion in this essay is limited to finite frames of discernment. In order to avoid the constant repetition of the hypothesis of finiteness, and even though most of the ideas we will encounter can be extended to infinite frames,* I ask the reader to assume that a frame of discernment is finite by definition.

§2. Basic Probability Numbers

Equipped with the idea of a frame of discernment, we can now give mathematical form to the intuitive picture wherein a portion of belief can be committed to a proposition but need not be committed either to it or to its negation.

A portion of belief committed to one proposition is thereby committed to any other proposition it implies. In terms of a frame of discernment, this means that a portion of belief committed to one subset is also committed to any subset containing it. So of the total belief committed to a given subset A of a frame Θ, some may also be committed to one or more proper subsets of A, while the rest will be committed exactly to A — to A and to no smaller subset.

*Most of the ideas of Chapters 2, 3, and 7 are extended to infinite frames in my *Allocations of Probability*.

It ought to be possible to partition all of one's belief among the different subsets of Θ, assigning to each subset A that portion that is committed to A and to nothing smaller. This suggests the following definition:

DEFINITION. If Θ is a frame of discernment, then a function $m: 2^{\Theta} \to [0,1]$ is called a *basic probability assignment* whenever

$$(1)\quad m(\emptyset) = 0$$

and

$$(2)\quad \sum_{A \subset \Theta} m(A) = 1 .$$

The quantity m(A) is called A's *basic probability number*, and it is understood to be the measure of the belief that is committed exactly to A. Condition (1) reflects the fact that no belief ought to be committed to $\emptyset$, while (2) reflects the convention that one's total belief has measure one.

To reiterate, the quantity m(A) measures the belief that one commits exactly to A, not the total belief that one commits to A. To obtain the measure of the total belief committed to A, one must add to m(A) the quantities m(B) for all proper subsets B of A:

$$\text{Bel}(A) = \sum_{B \subset A} m(B) . \tag{2.1}$$

A function $\text{Bel}: 2^{\Theta} \to [0,1]$ is called a *belief function* over Θ if it is given by (2.1) for some basic probability assignment $m: 2^{\Theta} \to [0,1]$.

The belief function with the simplest structure is surely the one obtained by setting $m(\Theta) = 1$ and $m(A) = 0$ for all $A \neq \Theta$; it has $\text{Bel}(\Theta) = 1$ but $\text{Bel}(A) = 0$ for all $A \neq \Theta$. Since this belief function seems appropriate when one has no evidence, it is called the *vacuous* belief function.*

*This name was suggested to me by John Tukey.

The reader for whom the ideas of this section are not transparent may wish to turn immediately to §1 of Chapter 4, where another simple example is exhibited.

§3. Belief Functions

The class of belief functions can be characterized without reference to basic probability assignments:

THEOREM 2.1.* *If* Θ *is a frame of discernment, then a function* $\mathrm{Bel}: 2^{\Theta} \to [0,1]$ *is a belief function if and only if it satisfies the following conditions:*

(1) $\mathrm{Bel}(\emptyset) = 0$.

(2) $\mathrm{Bel}(\Theta) = 1$.

(3) *For every positive integer* n *and every collection* $A_1, \cdots, A_n$ *of subsets of* Θ,

$$\mathrm{Bel}(A_1 \cup \cdots \cup A_n) \geq \sum_{\substack{I \subset \{1,\cdots,n\} \\ I \neq \emptyset}} (-1)^{|I|+1} \mathrm{Bel}\left(\bigcap_{i \in I} A_i\right).$$

Furthermore, the basic probability assignment that produces a given belief function is unique and can be recovered from the belief function:

THEOREM 2.2. *Suppose* $\mathrm{Bel}: 2^{\Theta} \to [0,1]$ *is the belief function given by the basic probability assignment* $m: 2^{\Theta} \to [0,1]$. *Then*

$$m(A) = \sum_{B \subset A} (-1)^{|A-B|} \mathrm{Bel}(B) \tag{2.2}$$

for all $A \subset \Theta$.

When discussing a belief function Bel in the sequel, I will often take for granted that its basic probability assignment is denoted by m.

---

*Displayed theorems are proven in the mathematical appendices to the various chapters.

A subset A of a frame Θ is called a *focal element* of a belief function Bel over Θ if $m(A) > 0$. The union of all the focal elements of a belief function is called its *core*.

THEOREM 2.3. *Suppose* $\mathcal{C}$ *is the core of a belief function* Bel *over* Θ. *Then a subset* $B \subset \Theta$ *satisfies* $\mathrm{Bel}(B) = 1$ *if and only if* $\mathcal{C} \subset B$.

§4. Commonality Numbers

The intuitive picture underlying belief functions can be seen most vividly if we represent the set Θ geometrically. For if we think of the elements of Θ as points, then we can think of our portions of belief as semi-mobile "probability masses," which can sometimes move from point to point but are restricted in that various of them are committed to, or confined to, various subsets of Θ.

The basic probability numbers are easily understood in terms of this image: $m(A)$ measures the total portion of belief, or the total probability mass, that is confined to A yet none of which is confined to any proper subset of A. In other words, $m(A)$ measures the probability mass that is confined to A but can move freely to every point of A.

The quantity

$$Q(A) = \sum_{\substack{B \subset \Theta \\ A \subset B}} m(B) \tag{2.3}$$

therefore measures the total probability mass that can move freely to every point of A. (Notice that $Q(\emptyset) = 1$.) I will call the quantity $Q(A)$ the *commonality number* for A, and I will call the function $Q : 2^{\Theta} \to [0, 1]$ the *commonality function* for Bel.*

---

*Arthur Dempster introduced the commonality numbers, though he did not coin a name for them. See pp. 335-337 of his 1967 paper.

The commonality function provides yet another way of specifying a belief function, for one can recover the belief function from it:

THEOREM 2.4. *Suppose* Bel *is a belief function over* Θ *and* Q *is its commonality function. Then*

$$\mathrm{Bel}(A) = \sum_{B \subset \bar{A}} (-1)^{|B|} Q(B) \tag{2.4}$$

and

$$Q(A) = \sum_{B \subset A} (-1)^{|B|} \mathrm{Bel}(\bar{B}) \tag{2.5}$$

for all $A \subset \Theta$.

The commonality numbers also provide another way of characterizing the core of a belief function:

THEOREM 2.5. *Suppose* $\mathcal{C}$ *is the core and* Q *is the commonality function of a belief function over* Θ. *Then an element* $\theta \in \Theta$ *is in* $\mathcal{C}$ *if and only if* $Q(\{\theta\}) > 0$.*

Since a commonality function is non-increasing ($B \subset A$ implies $Q(B) \geq Q(A)$), it follows from Theorem 2.5 that $Q(A) = 0$ whenever A includes a point not in the core. But some belief functions also award zero commonality numbers to subsets of their cores.

EXAMPLE 2.1. *A Belief Function Whose Core Has a Zero Commonality Number*. Set $\Theta = \{a, b, c, d\}$, and define a basic probability assignment $m: 2^{\Theta} \to [0,1]$ by $m(\{a,b\}) = \frac{1}{3}$, $m(\{c\}) = \frac{2}{3}$, and $m(A) = 0$ for all other $A \subset \Theta$. Then the core of the corresponding belief function is equal to $\{a, b, c\}$. Hence the singletons $\{a\}$,

*I will often use a lower case θ to denote a typical element of the set Θ – i.e., a typical value of the parameter $\boldsymbol{\theta}$.

$\{b\}$ and $\{c\}$ all have positive commonality numbers. But the subsets $\{a,c\}$, $\{b,c\}$ and $\{a,b,c\}$ all have zero commonality numbers, even though they are subsets of the core. ■

If we set $A = \emptyset$ in (2.4), we obtain

$$0 = \sum_{B \subset \Theta} (-1)^{|B|} Q(B) ,$$

or

$$\sum_{\substack{A \subset \Theta \\ A \neq \emptyset}} (-1)^{|A|+1} Q(A) = 1 ;$$

and it is evident from this formula that all the commonality numbers for a belief function are determined once the commonality numbers for non-empty subsets are determined modulo a common factor. More explicitly, if we knew that

$$Q(A) = Kq(A)$$

for all non-empty $A \subset \Theta$, where $q : (2^{\Theta} - \{\emptyset\}) \to [0, \infty)$ was a known function but K was an unknown positive constant, then we could calculate K;

$$\sum_{\substack{A \subset \Theta \\ A \neq \emptyset}} (-1)^{|A|+1} Kq(A) = 1 ,$$

or

$$K = \left(\sum_{\substack{A \subset \Theta \\ A \neq \emptyset}} (-1)^{|A|+1} q(A) \right)^{-1} . \tag{2.6}$$

And we could then calculate the values of Q.

§5. Degrees of Doubt and Upper Probabilities

One's beliefs about a proposition A are not fully described by one's degree of belief $\text{Bel}(A)$, for $\text{Bel}(A)$ does not reveal to what extent one doubts A – i.e., to what extent one believes its negation $\bar{A}$. A fuller description consists of the degree of belief Bel(A) together with the *degree of doubt*

$$\mathrm{Dou}(A) = \mathrm{Bel}(\bar{A}) .$$

The degree of doubt is less useful, though, than the quantity

$$P^*(A) = 1 - \mathrm{Dou}(A) ,$$

which expresses the extent to which one fails to doubt A – i.e., the extent to which one finds A credible or plausible. For lack of a more expressive name, I call $P^*(A)$ the *upper probability* of A.

Whenever Bel is belief function over a frame Θ, the function $P^* : 2^{\Theta} \to [0,1]$ defined by

$$P^*(A) = 1 - \mathrm{Bel}(\bar{A}) \tag{2.7}$$

is called the *upper probability function* for Bel. Since

$$\mathrm{Bel}(A) = 1 - P^*(\bar{A})$$

for all $A \subset \Theta$, the functions Bel and P^* convey precisely the same information; either may be obtained from the other.

Using (2.7), we can express $P^*(A)$ in terms of Bel's basic probability assignment m:

$$\begin{aligned} P^*(A) = 1 - \mathrm{Bel}(\bar{A}) &= \sum_{B \subset \Theta} m(B) - \sum_{B \subset \bar{A}} m(B) \\ &= \sum_{B \cap A \neq \emptyset} m(B) . \end{aligned} \tag{2.8}$$

This formula affords us a neat interpretation of $P^*(A)$ in terms of our geometric picture: since the probability mass measured by m(B) can move into A if and only if $B \cap A \neq \emptyset$, $P^*(A)$ measures the total probability mass that can move into A.

When we compare (2.8) with (2.1), we immediately notice that

$$\mathrm{Bel}(A) \leq P^*(A) ;$$

the probability mass constrained to A is included in the total probability mass that can move into A. And when we compare (2.8) with (2.2), we notice that

$$P^*(\{\theta\}) = Q(\{\theta\})$$

for any particular element θ of Θ; the probability mass that can move into a singleton is the same as the probability mass that can move to every point of it.

The relation between the upper probabilities and commonality numbers for larger subsets is more complicated but equally symmetric:

THEOREM 2.6. *Suppose* P^* *and* Q *are the upper probability function and the commonality function for some belief function over* Θ. *Then*

$$P^*(A) = \sum_{\substack{B \subset A \\ B \neq \emptyset}} (-1)^{|B|+1} Q(B) \tag{2.9}$$

and

$$Q(A) = \sum_{B \subset A} (-1)^{|B|+1} P^*(B) \tag{2.10}$$

for all non-empty $A \subset \Theta$.

§6. Bayesian Belief Functions

The first three of Bayes' rules can also be expressed in terms of a frame of discernment:

DEFINITION. If Θ is a frame of discernment, then a function $\mathrm{Bel}: 2^\Theta \to [0,1]$ is called a *Bayesian belief function* if

(1) $\mathrm{Bel}(\emptyset) = 0$,

(2) $\mathrm{Bel}(\Theta) = 1$,

(3) $\mathrm{Bel}(A \cup B) = \mathrm{Bel}(A) + \mathrm{Bel}(B)$ whenever $A, B \subset \Theta$ and $A \cap B = \emptyset$.

THEOREM 2.7. *A Bayesian belief function is a belief function.*

THEOREM 2.8. *Suppose* $\mathrm{Bel}: 2^{\Theta} \to [0,1]$ *is a belief function, with upper probability function* P^*. *Then the following assertions are all equivalent:*

(1) Bel *is Bayesian.*

(2) *All of* Bel's *focal elements are singletons.*

(3) Bel *awards a zero commonality number to any subset containing more than one element.*

(4) $\mathrm{Bel} = P^*$.

(5) $\mathrm{Bel}(A) + \mathrm{Bel}(\bar{A}) = 1$ *for all* $A \subset \Theta$.

When a focal element of a belief function consists of a single point, the corresponding probability mass is constrained to stay at that point and has no freedom of movement. So (2) of Theorem 2.8 means that in the geometric picture associated with a Bayesian belief function none of the probability mass has any freedom of movement.

Bayesian belief functions are clearly simpler and easier to describe than belief functions in general. As (2) of Theorem 2.8 indicates and Theorem 2.9 makes explicit, a Bayesian belief function is more like a point function than a set function in its level of complexity.

THEOREM 2.9. *A function* $\mathrm{Bel}: 2^{\Theta} \to [0,1]$ *is a Bayesian belief function if and only if there exists a function* $p: \Theta \to [0,1]$ *such that*

$$\sum_{\theta \in \Theta} p(\theta) = 1$$

and

$$\mathrm{Bel}(A) = \sum_{\theta \in A} p(\theta)$$

for all $A \subset \Theta$. *(If* Bel *is a Bayesian belief function, then the function* p *is of course unique; it is given by* $p(\theta) = m(\{\theta\})$.)

But as I have already argued, and as we will see more clearly in Chapter 9, this simplicity renders the Bayesian belief functions awkward for the representation of evidence.

§7. Mathematical Appendix

This appendix begins with a review of notation and then turns to proving the chapter's displayed theorems.

* *

Set Notation. A set is often denoted by listing its elements within brackets. For example, if $a_1, \cdots, a_n$ are the elements of a set A, then we write $A = \{a_1, \cdots, a_n\}$. In particular, we write $\{a\}$ for the set that contains only the single element a. The set of objects a satisfying a certain condition is often denoted $\{a | \cdots\}$, the condition taking the place of the dots.

The set-theoretic symbols ϵ and $\subset$ are undoubtedly familiar to the reader; $a \epsilon A$ means that a is an element of A, while $A \subset B$ means that A is a subset of B. These symbols are negated by a slash: $a \notin A$, $A \not\subset B$.

Suppose A and B are sets. Then $A \cap B$ denotes the intersection of A and B, or the set of all elements that are in both sets; $A \cup B$ denotes the union of A and B, or the set of all elements that are in at least one of the two sets; and $A - B$ denotes the difference of A and B, or the set of all the elements of A that are not in B. When studying the subsets of a particular set Θ, one often uses the symbol $\bar{A}$ to denote $\Theta - A$.

The notations for union and intersection are easily extended to collections of sets. For example, $A_1 \cup \cdots \cup A_n$ denotes the union of the sets $A_1, \cdots, A_n$. And $\bigcap_{i \epsilon I} A_i$ denotes the intersection of those A_i for which $i \epsilon I$.

The symbol $|A|$ denotes the *cardinality* of the set A – i.e., the number of elements in A. Notice that $(-1)^{|A|}$ is $+1$ if the cardinality of A is even, -1 if it is odd; this number is called the *parity* of A. Notice also that if $B \subset A$, then $|A-B| = |A| - |B|$ and $(-1)^{|A-B|} = (-1)^{|A|}(-1)^{|B|}$.

Special Sets. The symbol $\emptyset$ denotes the empty set – i.e., the set with no elements. We also use the usual notation for intervals of real numbers: $[0,1]$ is the interval of real numbers from zero to one, inclusively; $[0,\infty)$ is the set of all non-negative real numbers; and $[0,\infty]$ is the set of all non-negative real numbers together with infinity.

Mappings and Functions. The notation $f: A \to B$ indicates that f is a *mapping*, assigning to each element a of A an element $f(a)$ of B. When B consists of real numbers, such a mapping is called a *function*.

A mapping $f: A \to B$ is *one-to-one* if the elements $f(a_1)$ and $f(a_2)$ of B are distinct whenever the elements a_1 and a_2 of A are distinct. It is *onto* if for every $b \in B$ there exists $a \in A$ such that $f(a) = b$.

We use the usual symbols for the exponential function: $e^x = \exp(x)$ is the number e raised to the power x. And $\log(x)$ is used to denote the natural logarithm of x – i.e., the logarithm to the base e.

Summations. The first symbol under a summation sign is usually its index of summation; the conditions that follow indicate the range of the index.

*   *

LEMMA 2.1. *If* A *is a finite set, then*

$$\sum_{B \subset A} (-1)^{|B|} = \begin{cases} 1 & \textit{if } A = \emptyset \\ 0 & \textit{otherwise.} \end{cases}$$

Proof of Lemma 2.1. By the binomial theorem,*

$$\binom{n}{0} - \binom{n}{1} + \binom{n}{2} - + \cdots + (-1)^n \binom{n}{n} = (1-1)^n = 0$$

whenever n is a positive integer. It follows that whenever $A = \{\theta_1, \cdots, \theta_n\}$ is a finite non-empty set,

---

*See, for example, p. 201 of Courant's calculus text.

$$\sum_{B\subset A} (-1)^{|B|} = (-1)^{|\emptyset|} + \sum_{i} (-1)^{|\{\theta_i\}|} + \sum_{i<j} (-1)^{|\{\theta_i,\theta_j\}|} + \cdots + (-1)^{|A|}$$

$$= \binom{n}{0} - \binom{n}{1} + \binom{n}{2} - + \cdots + (-1)^n \binom{n}{n} = 0 .$$

And, of course, when A is empty,

$$\sum_{B\subset A} (-1)^{|B|} = (-1)^{|A|} = (-1)^0 = 1 . \blacksquare$$

LEMMA 2.2. *If* A *is a finite set and* $B\subset A$, *then*

$$\sum_{\substack{C \\ B\subset C\subset A}} (-1)^{|C|} = \begin{cases} (-1)^{|A|} & \textit{if } A = B \\ 0 & \textit{otherwise.} \end{cases}$$

Proof of Lemma 2.2. This lemma is seen to follow from Lemma 2.1 once it is noticed that

$$\sum_{\substack{C \\ B\subset C\subset A}} (-1)^{|C|} = \sum_{D\subset(A-B)} (-1)^{|B\cup D|} = (-1)^{|B|} \sum_{D\subset(A-B)} (-1)^{|D|} . \blacksquare$$

LEMMA 2.3.* *Suppose* Θ *is a finite set and* f *and* g *are functions on* 2^Θ. *Then*

$$f(A) = \sum_{B\subset A} g(B) \tag{2.11}$$

for all $A\subset\Theta$ *if and only if*

$$g(A) = \sum_{B\subset A} (-1)^{|A-B|} f(B) \tag{2.12}$$

for all $A\subset\Theta$.

---

*Lemmas 2.3 and 2.4 are examples of a general combinatorial phenomena that C.-G. Rota christened *Möbius inversion*. See p. 15 of Hall's *Combinatorial Theory*.

Proof of Lemma 2.3. Both implications follow by simple calculations using Lemma 2.2.

(i) If (2.11) holds for all $A \subset \Theta$, then

$$\sum_{B \subset A} (-1)^{|A-B|} f(B) = (-1)^{|A|} \sum_{B \subset A} (-1)^{|B|} f(B)$$

$$= (-1)^{|A|} \sum_{B \subset A} (-1)^{|B|} \sum_{C \subset B} g(C)$$

$$= (-1)^{|A|} \sum_{C \subset A} g(C) \sum_{\substack{B \\ C \subset B \subset A}} (-1)^{|B|}$$

$$= (-1)^{|A|} g(A)(-1)^{|A|}$$

$$= g(A) .$$

(ii) If (2.12) holds for all $A \subset \Theta$, then

$$\sum_{B \subset A} g(B) = \sum_{B \subset A} \sum_{C \subset B} (-1)^{|B-C|} f(C)$$

$$= \sum_{C \subset A} (-1)^{|C|} f(C) \sum_{\substack{B \\ C \subset B \subset A}} (-1)^{|B|}$$

$$= (-1)^{|A|} f(A)(-1)^{|A|}$$

$$= f(A) . \blacksquare$$

LEMMA 2.4. *Suppose* Θ *is a finite set and* f *and* g *are functions on* 2^{Θ}. *Then*

$$f(A) = \sum_{B \subset A} (-1)^{|B|+1} g(B) \tag{2.13}$$

for all $A \subset \Theta$ *if and only if*

$$g(A) = \sum_{B \subset A} (-1)^{|B|+1} f(B)$$

for all $A \subset \Theta$.

Proof of Lemma 2.4. If (2.13) holds for all $A \subset \Theta$, then

$$\sum_{B \subset A} (-1)^{|B|+1} f(B) = \sum_{B \subset A} (-1)^{|B|+1} \sum_{C \subset B} (-1)^{|C|+1} g(C)$$

$$= \sum_{C \subset A} (-1)^{|C|} g(C) \sum_{\substack{B \\ C \subset B \subset A}} (-1)^{|B|}$$

$$= g(A) . \blacksquare$$

LEMMA 2.5. *Suppose* Θ *is a finite set and* f *and* g *are functions on* 2^Θ. *Then*

$$f(A) = \sum_{B \subset \bar{A}} (-1)^{|B|} g(B) \tag{2.14}$$

for all $A \subset \Theta$ *if and only if*

$$g(A) = \sum_{B \subset A} (-1)^{|B|} f(\bar{B}) \tag{2.15}$$

for all $A \subset \Theta$.

Proof of Lemma 2.5. Let h denote the function on 2^Θ given by $h(A) = -f(\bar{A})$ for all $A \subset \Theta$.

(i) Suppose (2.14) holds for all $A \subset \Theta$. Then

$$h(A) = \sum_{B \subset A} (-1)^{|B|+1} g(B) , \tag{2.16}$$

and by Lemma 2.4,

$$g(A) = \sum_{B \subset A} (-1)^{|B|+1} h(B) \tag{2.17}$$

$$= \sum_{B \subset A} (-1)^{|B|} f(\bar{B})$$

for all $A \subset \Theta$.

(ii) Suppose (2.15) holds for all $A \subset \Theta$. Then (2.17) holds, and Lemma 2.4 yields (2.16) and thus (2.14). ■

Proof of Theorem 2.1.

(i) If Bel is given by (2.1) for some basic probability assignment m, then conditions (1) and (2) of the theorem follow immediately from conditions (1) and (2) of the definition of a basic probability assignment. To show that (3) holds, fix a collection $A_1, \cdots, A_n$ of subsets of Θ, and set $I(B) = \{i \mid 1 \le i \le n;\ B \subset A_i\}$ for each $B \subset \Theta$. Using Lemma 2.1, we find that

$$\sum_{\substack{I \subset \{1,\cdots,n\} \\ I \neq \emptyset}} (-1)^{|I|+1} \mathrm{Bel}\Big(\bigcap_{i \in I} A_i\Big) = \sum_{\substack{I \subset \{1,\cdots,n\} \\ I \neq \emptyset}} (-1)^{|I|+1} \sum_{B \subset \bigcap_{i \in I} A_i} m(B)$$

$$= \sum_{\substack{B \subset \Theta \\ I(B) \neq \emptyset}} m(B) \sum_{\substack{I \subset I(B) \\ I \neq \emptyset}} (-1)^{|I|+1}$$

$$= \sum_{\substack{B \subset \Theta \\ I(B) \neq \emptyset}} m(B) \Big(1 - \sum_{I \subset I(B)} (-1)^{|I|}\Big)$$

$$= \sum_{\substack{B \subset \Theta \\ I(B) \neq \emptyset}} m(B) = \sum_{\substack{B \subset \Theta \\ B \subset A_i \text{ for some } i}} m(B) ,$$

and this is certainly less than or equal to

$$\mathrm{Bel}(A_1 \cup \cdots \cup A_n) = \sum_{B \subset A_1 \cup \cdots \cup A_n} m(B) .$$

(ii) Consider a function $\mathrm{Bel}: 2^\Theta \to [0,1]$ satisfying the three conditions of the theorem, and define a function m on 2^Θ by (2.2). Then by Lemma 2.3, Bel can be expressed in terms of m by (2.1). So to show that Bel is a belief function, we need only show that m is a basic probability assignment – i.e., that $m(\emptyset) = 0$, $\sum_{A \subset \Theta} m(A) = 1$, and $m(A) \ge 0$ for all $A \subset \Theta$.

It follows from (2.2) and (1) that $m(\emptyset) = 0$, and it follows from (2.1) and (2) that $\sum_{A \subset \Theta} m(A) = 1$. To see that $m(A) \ge 0$ whenever $A \neq \emptyset$, set

$A = \{\theta_1, \cdots, \theta_n\}$, where $n \geq 1$ and the θ_i are distinct. And set $A_i = A - \{\theta_i\}$, so that $A_1, \cdots, A_n$ are precisely the subsets of A that omit exactly one element of A. Then every proper subset B of A can be uniquely expressed as an intersection of the A_i; indeed, if $A - B = \{\theta_{i_1}, \cdots, \theta_{i_k}\}$, then $B = A_{i_1} \cap \cdots \cap A_{i_k}$. Thus

$$m(A) = \sum_{B \subset A} (-1)^{|A-B|} \, Bel(B)$$

$$= Bel(A) + \sum_{\substack{I \subset \{1,\cdots,n\} \\ I \neq \emptyset}} (-1)^{|I|} \, Bel\left(\bigcap_{i \in I} A_i\right)$$

$$= Bel(A) - \sum_{\substack{I \subset \{1,\cdots,n\} \\ I \neq \emptyset}} (-1)^{|I|+1} \, Bel\left(\bigcap_{i \in I} A_i\right),$$

and this is non-negative by (3). ■

Proof of Theorem 2.2. Theorem 2.2 is an immediate consequence of Lemma 2.3. ■

Proof of Theorem 2.3. A subset $B \subset \Theta$ will satisfy

$$Bel(B) = \sum_{C \subset B} m(C) = 1$$

if and only if $m(C) = 0$ for all C not contained in B. This is equivalent to requiring all the focal elements to be contained in B. And this in turn is equivalent to their union, the core, being contained in B. ■

Proof of Theorem 2.4. Using (2.3) and Lemma 2.1, we find that

$$\sum_{B \subset \bar{A}} (-1)^{|B|} Q(B) = \sum_{B \subset \bar{A}} (-1)^{|B|} \sum_{\substack{C \\ B \subset C}} m(C)$$

$$= \sum_{C \subset \Theta} m(C) \sum_{B \subset C \cap \bar{A}} (-1)^{|B|}$$

$$= \sum_{\substack{C \\ C \cap \bar{A} = \emptyset}} m(C) = \sum_{C \subset A} m(C)$$

$$= \mathrm{Bel}(A)$$

for all $A \subset \Theta$. (2.5) then follows by Lemma 2.5. ■

Proof of Theorem 2.5. Since

$$Q(\{\theta\}) = \sum_{\substack{B \subset \Theta \\ \theta \in B}} m(B) \ ,$$

$Q(\{\theta\})$ will be positive if and only if θ is in at least one of the focal elements – i.e., if and only if θ is in the core. ■

Proof of Theorem 2.6. Using Theorem 2.4, we find that

$$P^*(A) = 1 - \mathrm{Bel}(\bar{A}) = 1 - \sum_{B \subset A} (-1)^{|B|} Q(B)$$

$$= \sum_{\substack{B \subset A \\ B \neq \emptyset}} (-1)^{|B|+1} Q(B)$$

for all $A \subset \Theta$. The second relation then follows by Lemma 2.4. ■

LEMMA 2.6. *A Bayesian belief function* $\mathrm{Bel}: 2^\Theta \to [0,1]$ *satisfies*

$$\mathrm{Bel}(A) = \sum_{\theta \in A} \mathrm{Bel}(\{\theta\})$$

for all $A \subset \Theta$.

Proof of Lemma 2.6. The lemma is true by convention if $A = \emptyset$. If $A \neq \emptyset$, then we may write $A = \{\theta_1, \cdots, \theta_n\}$, where the θ_i are distinct. And then repeated application of the third rule in the definition of Bayesian belief functions gives

$$\begin{aligned} \mathrm{Bel}(A) &= \mathrm{Bel}(\{\theta_1\}) + \mathrm{Bel}(\{\theta_2, \cdots, \theta_n\}) \\ &= \mathrm{Bel}(\{\theta_1\}) + \mathrm{Bel}(\{\theta_2\}) + \mathrm{Bel}(\{\theta_3, \cdots, \theta_n\}) \\ &= \cdots = \mathrm{Bel}(\{\theta_1\}) + \mathrm{Bel}(\{\theta_2\}) + \cdots + \mathrm{Bel}(\{\theta_n\}) . \ \blacksquare \end{aligned}$$

Proof of Theorem 2.7. Suppose $\mathrm{Bel}: 2^\Theta \to [0,1]$ is a Bayesian belief function. Define $m: 2^\Theta \to [0,1]$ by $m(\{\theta\}) = \mathrm{Bel}(\{\theta\})$ for all $\theta \in \Theta$ and $m(A) = 0$ for all non-singleton subsets A of Θ. Then $m(\emptyset) = 0$, and by Lemma 2.6,

$$\sum_{A \subset \Theta} m(A) = \sum_{\theta \in \Theta} m(\{\theta\}) = \sum_{\theta \in \Theta} \mathrm{Bel}(\{\theta\}) = \mathrm{Bel}(\Theta) = 1 \ ;$$

so m is a basic probability assignment. And for any $A \subset \Theta$,

$$\sum_{B \subset A} m(B) = \sum_{\theta \in A} m(\{\theta\}) = \sum_{\theta \in A} \mathrm{Bel}(\{\theta\}) = \mathrm{Bel}(A) \ ;$$

so Bel is the belief function given by m. ■

LEMMA 2.7. *A belief function* $\mathrm{Bel}: 2^\Theta \to [0,1]$ *is Bayesian if and only if its basic probability assignment* m *is given by* $m(\{\theta\}) = \mathrm{Bel}(\{\theta\})$ *and* $m(A) = 0$ *for all non-singleton subsets* A *of* Θ.

Proof of Lemma 2.7. If Bel is Bayesian, then we know from the preceding proof that m is given in this way. If m is given in this way, then

$$\begin{aligned}\mathrm{Bel}(A) + \mathrm{Bel}(B) &= \sum_{C \subset A} m(C) + \sum_{C \subset B} m(C) \\ &= \sum_{\theta \in A} m(\{\theta\}) + \sum_{\theta \in B} m(\{\theta\}) \\ &= \sum_{\theta \in A \cup B} m(\{\theta\}) = \sum_{C \subset A \cup B} m(C) \\ &= \mathrm{Bel}(A \cup B)\end{aligned}$$

whenever $A, B \subset \Theta$ and $A \cap B = \emptyset$; and thus Bel is Bayesian. ∎

Proof of Theorem 2.8.

(i) The equivalence of (1) and (2) follows from Lemma 2.7.

(ii) The equivalence of (2) and (3) follows from the relation

$$Q(A) = \sum_{\substack{B \\ A \subset B}} m(B) .$$

(iii) The equivalence of (2) and (4) follows from a comparison of the relations

$$\mathrm{Bel}(A) = \sum_{B \subset A} m(B) \qquad \text{and} \qquad P^*(A) = \sum_{B \cap A \neq \emptyset} m(B) .$$

(iv) The equivalence of (4) and (5) follows from the relation

$$P^*(A) = 1 - \mathrm{Bel}(\bar{A}) . \;∎$$

Proof of Theorem 2.9.

(i) Suppose there exists a function $p: \Theta \to [0,1]$ satisfying the conditions of the theorem. Then

$$\mathrm{Bel}(\emptyset) = \sum_{\theta \in \emptyset} p(\theta) = 0 ,$$

and

$$\text{Bel}(\Theta) = \sum_{\theta \in \Theta} p(\theta) = 1 .$$

And whenever A and B are disjoint subsets of Θ,

$$\text{Bel}(A) + \text{Bel}(B) = \sum_{\theta \in A} p(\theta) + \sum_{\theta \in B} p(\theta) = \sum_{\theta \in A \cup B} p(\theta)$$

$$= \text{Bel}(A \cup B) .$$

So Bel is a Bayesian belief function.

(ii) Suppose Bel is a Bayesian belief function. Then by Lemma 2.7, the function $p : \Theta \to [0, 1]$ defined by $p(\theta) = m(\{\theta\})$ satisfies the conditions of the theorem.

(iii) The function p satisfying the conditions of the theorem must be unique, for it must satisfy $p(\theta) = \text{Bel}(\{\theta\}) = m(\{\theta\})$. ∎

CHAPTER 3. DEMPSTER'S RULE OF COMBINATION

> I think I perceive or remember something but am not sure; this would seem to give me some ground for believing it, contrary to Mr. Keynes' theory, by which the degree of belief in it which it would be rational for me to have is that given by the probability relation between the proposition in question and the things I know for certain.
>
> FRANK PLUMPTON RAMSEY (1903-1930)

Belief functions are adapted to the representation of evidence because they admit a genuine rule of combination. Given several belief functions over the same frame of discernment but based on distinct bodies of evidence, *Dempster's rule of combination* enables us to compute their *orthogonal sum*, a new belief function based on the combined evidence. Though this essay provides no conclusive *a priori* argument for Dempster's rule, we will see in the following chapters that the rule does seem to reflect the pooling of evidence, provided only that the belief functions to be combined are actually based on entirely distinct bodies of evidence and that the frame of discernment discerns the relevant interaction of those bodies of evidence.*

In the special case of a frame of discernment containing only two elements, Dempster's rule of combination was accurately stated and used by Johann Heinrich Lambert in his *Neues Organon*, published in 1764.** But the general formulation was not achieved until two hundred years later; it was first published by Arthur Dempster in 1967.***

---

*The condition that the frame discern relevant interaction is explained in §2 of Chapter 8.

**See pp. 318-421 of Volume 2. This volume was reprinted in 1965 as the second volume of Lambert's *Philosophische Schriften*.

***See pp. 335-337 of Dempster's 1967 article.

§1. Combining Two Belief Functions

Dempster's rule is most accessible to the intuition when it is expressed in terms of the basic probability numbers, and especially when these basic probability numbers are depicted geometrically.

Suppose m_1 is the basic probability assignment for a belief function Bel_1 over a frame Θ, and denote Bel_1's focal elements by $A_1,\cdots,A_k$. Then the probability masses measured by the basic probability numbers $m_1(A_1),\cdots,m_1(A_k)$ can be depicted as segments of a line segment of length one, as in Figure 3.1. Bear in mind that this is a picture of one's probability masses, not of the frame of discernment.

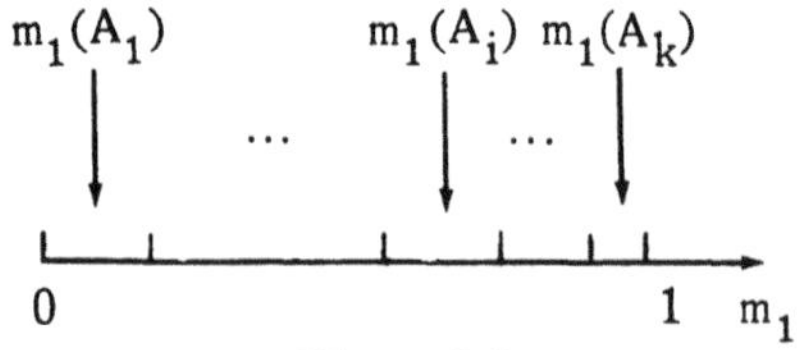

Figure 3.1

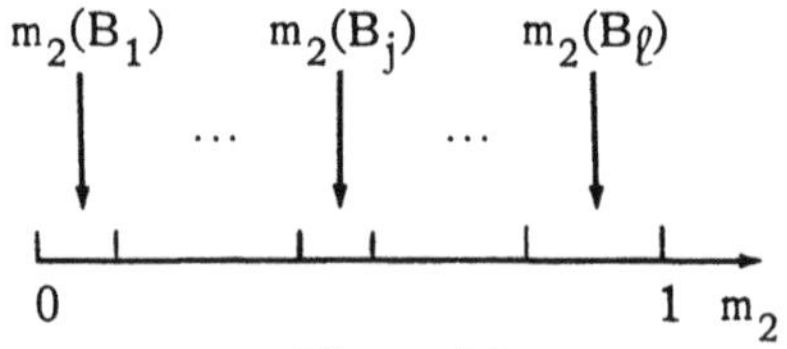

Figure 3.2

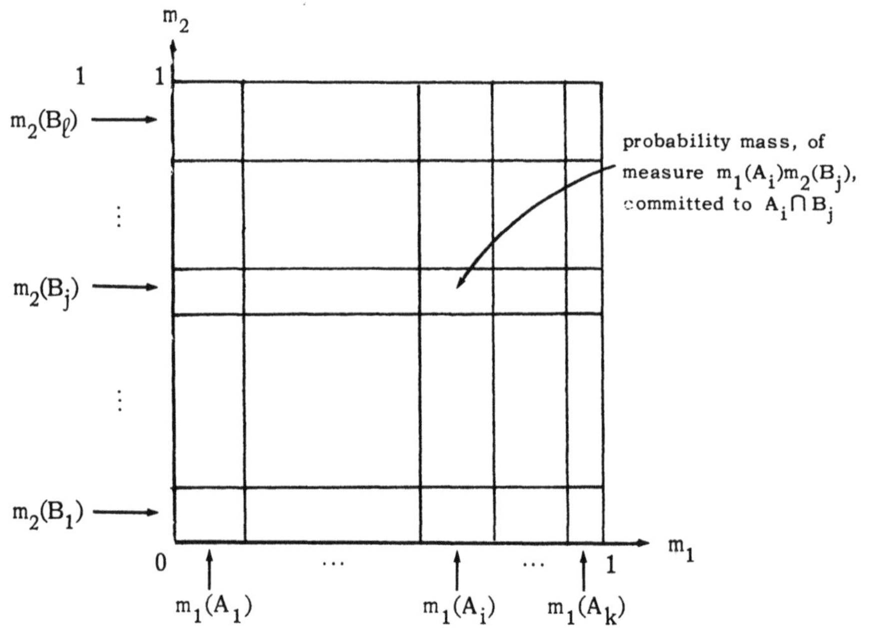

Figure 3.3

Figure 3.2 depicts in the same way the basic probability numbers for a second belief function Bel_2 with basic probability assignment m_2 and focal elements $B_1, \cdots, B_\ell$. And Figure 3.3 shows how the two line segments representing m_1 and m_2 can be orthogonally combined to obtain a square.

In order to carry out the combination of Bel_1 and Bel_2, we now think of the square as representing our total probability mass, and suppose that Bel_1 commits vertical strips to its focal elements, while Bel_2 commits horizontal strips to its focal elements. The figure singles out, for example, a vertical strip of measure $m_1(A_i)$ that is exactly committed to A_i by m_1 and a horizontal strip of measure $m_2(B_j)$ that is exactly committed to B_j by m_2. The intersection of these two strips has measure $m_1(A_i)m_2(B_j)$, and since it is committed both to A_i and to B_j, we may say that the joint effect of Bel_1 and Bel_2 is to commit it exactly to $A_i \cap B_j$.

Similarly, we can specify the exact commitment of every rectangle in the figure. A given subset A of Θ may have more than one of these rectangles exactly committed to it, of course; the total probability mass exactly committed to A will have measure

$$\sum_{\substack{i,j \\ A_i \cap B_j = A}} m_1(A_i)m_2(B_j) .$$

The only difficulty with this scheme is that it may commit some of the square to the empty set $\emptyset$. For there may well be a focal element A_i of Bel_1 and a focal element B_j of Bel_2 such that $A_i \cap B_j = \emptyset$, in which case

$$\sum_{\substack{i,j \\ A_i \cap B_j = \emptyset}} m_1(A_i)m_2(B_j) > 0 .$$

The only remedy is to "discard" all the rectangles thus committed to $\emptyset$. If not all the rectangles are thus discarded, the measures of the remaining rectangles can then be inflated by multiplying them by the factor

$$\left(1 - \sum_{\substack{i,j \\ A_i \cap B_j = \emptyset}} m_1(A_i)\, m_2(B_j)\right)^{-1} \tag{3.1}$$

so that the total probability mass will again have measure one.

As the next two theorems show, this construction does indeed lead to a new basic probability assignment, provided only that Bel_1 and Bel_2 do not flatly contradict each other.

THEOREM 3.1. *Suppose* Bel_1 *and* Bel_2 *are belief functions over the same frame* Θ, *with basic probability assignments* m_1 *and* m_2 *and focal elements* $A_1, \cdots, A_k$ *and* $B_1, \cdots, B_\ell$, *respectively. Suppose*

$$\sum_{\substack{i,j \\ A_i \cap B_j = \emptyset}} m_1(A_i)\, m_2(B_j) < 1 \,. \tag{3.2}$$

Then the function $m : 2^\Theta \to [0,1]$ *defined by* $m(\emptyset) = 0$ *and*

$$m(A) = \frac{\displaystyle\sum_{\substack{i,j \\ A_i \cap B_j = A}} m_1(A_i)\, m_2(B_j)}{1 - \displaystyle\sum_{\substack{i,j \\ A_i \cap B_j = \emptyset}} m_1(A_i)\, m_2(B_j)} \tag{3.3}$$

for all non-empty $A \subset \Theta$ *is a basic probability assignment. The core of the belief function given by* m *is equal to the intersection of the cores of* Bel_1 *and* Bel_2.

The belief function given by m is called the *orthogonal sum* of Bel_1 and Bel_2 and is denoted $\mathrm{Bel}_1 \oplus \mathrm{Bel}_2$. If (3.2) does not hold, then we say that the orthogonal sum $\mathrm{Bel}_1 \oplus \mathrm{Bel}_2$ does not exist.

THEOREM 3.2. *Suppose* Bel_1 *and* Bel_2 *are belief functions over the same frame* Θ, *and let* Q_1 *and* Q_2 *denote their commonality functions. Then the following conditions are all equivalent:*

(1) $\mathrm{Bel}_1 \oplus \mathrm{Bel}_2$ *does not exist.*

(2) *The cores of* Bel_1 *and* Bel_2 *are disjoint.*

(3) *There exists a subset* $A \subset \Theta$ *such that* $\mathrm{Bel}_1(A) = 1$ *and* $\mathrm{Bel}_2(\bar{A}) = 1$.

(4) $Q_1(A)Q_2(A) = 0$ *for all non-empty* $A \subset \Theta$.

§2. Multiplying Commonality Numbers

Though the construction of the preceding section may provide greater insight, Dempster's rule is most easily expressed in terms of the commonality numbers.

THEOREM 3.3. *Let* Bel_1 *and* Bel_2 *be two belief functions over* Θ, *and suppose* $\mathrm{Bel}_1 \oplus \mathrm{Bel}_2$ *exists. Denote by* Q_1, Q_2 *and* Q *the commonality functions for* Bel_1, Bel_2 *and* $\mathrm{Bel}_1 \oplus \mathrm{Bel}_2$, *respectively. Then*

$$Q(A) = KQ_1(A)Q_2(A)$$

for all non-empty $A \subset \Theta$, *the constant* K *being independent of* A.

According to (2.6),

$$K = \left(\sum_{\substack{B \subset \Theta \\ B \neq \emptyset}} (-1)^{|B|+1} Q_1(B) Q_2(B) \right)^{-1} . \tag{3.4}$$

And the proof of Theorem 3.3 reveals that K is the same renormalizing constant as (3.1):

$$K = \left(1 - \sum_{\substack{A,B \\ A \cap B = \emptyset}} m_1(A) m_2(B)\right)^{-1}$$

$$= \left(\sum_{\substack{A,B \\ A \cap B \neq \emptyset}} m_1(A) m_2(B)\right)^{-1},$$

where m_1 and m_2 are the basic probability functions for Q_1 and Q_2.

§3. Combining Several Belief Functions

The rule of combination described in the preceding sections is a rule for combining a pair of belief functions, but by repeatedly applying it one can obviously combine any number of belief functions. Indeed, in order to combine a collection $Bel_1, \cdots, Bel_n$ of belief functions, one can form the pairwise orthogonal sums

$$Bel_1 \oplus Bel_2,$$

$$(Bel_1 \oplus Bel_2) \oplus Bel_3,$$

$$((Bel_1 \oplus Bel_2) \oplus Bel_3) \oplus Bel_4,$$

etc. – continuing until all the Bel_i are included. If Dempster's rule meets its purpose, then each stage of this process should correspond to the addition of the evidence underlying the Bel_i entered at that stage, and the belief function Bel finally issuing from the process should represent the pooled evidence from all the Bel_i.

Now the body of evidence obtained by pooling several bodies of evidence will not depend on the order in which the pooling is done, and hence our equation of Dempster's rule with the pooling of evidence will be tenable only if the belief function Bel resulting from the pairwise combination of the Bel_i remains the same when the Bel_i are entered in a different order. Fortunately, Theorem 3.3 makes it obvious that this condition is satisfied. Indeed, if the pairwise orthogonal sum exists at each stage, and if we let $Q, Q_1, \cdots, Q_n$ denote the commonality functions for $\mathrm{Bel}, \mathrm{Bel}_1, \cdots, \mathrm{Bel}_n$, then Theorem 3.3 implies that

$$Q(A) = KQ_1(A) \cdots Q_n(A) \tag{3.5}$$

for all non-empty $A \subset \Theta$, where K is the product of the constants (3.4) obtained at each stage. Using (2.6), we may express K directly as

$$K = \left(\sum_{\substack{B \subset \Theta \\ B \neq \emptyset}} (-1)^{|B|+1} Q_1(B) \cdots Q_n(B) \right)^{-1} . \tag{3.6}$$

And both (3.5) and (3.6) are obviously unaffected by any permutation of the Bel_i.

Theorem 3.4 tells us just when a collection of belief functions can be combined.

THEOREM 3.4. *Suppose* $\mathrm{Bel}_1, \cdots, \mathrm{Bel}_n$ *are belief functions over the same frame* Θ. *Denote their cores by* $\mathcal{C}_1, \cdots, \mathcal{C}_n$ *and their commonality functions by* $Q_1, \cdots, Q_n$, *respectively. Then the following conditions are all equivalent:*

(1) $\mathcal{C}_1 \cap \cdots \cap \mathcal{C}_n = \emptyset$.

(2) $Q_1(A) \cdots Q_n(A) = 0$ *for all non-empty* $A \subset \Theta$.

(3) *There exist* $A_1, \cdots, A_n$ *of* Θ *such that* $A_1 \cap \cdots \cap A_n = \emptyset$ *but* $\mathrm{Bel}_i(A_i) = 1$ *for* $i = 1, \cdots, n$.

(4) *There exist subsets* A *and* B *of* Θ *such that* $A \cap B = \emptyset$ *yet*

$$A = E_1 \cap \cdots \cap E_r$$

and

$$B = F_1 \cap \cdots \cap F_s$$

for some subsets $E_1, \cdots, E_r, F_1, \cdots, F_s$ *each of which is assigned degree of belief one by at least one of the* Bel_i.

(5) *An attempt to form the orthogonal sums*

$$\mathrm{Bel}_1 \oplus \mathrm{Bel}_2$$

$$(\mathrm{Bel}_1 \oplus \mathrm{Bel}_2) \oplus \mathrm{Bel}_3$$

$$((\mathrm{Bel}_1 \oplus \mathrm{Bel}_2) \oplus \mathrm{Bel}_3) \oplus \mathrm{Bel}_4$$

$$((\cdots(\mathrm{Bel}_1 \oplus \mathrm{Bel}_2)\cdots) \oplus \mathrm{Bel}_{n-1}) \oplus \mathrm{Bel}_n$$

will fail; one will encounter an orthogonal sum that does not exist before one completes the list.

(6) *An attempt to combine the* Bel_i *iteratively in any other order will similarly fail.*

A collection of belief functions $\mathrm{Bel}_1, \cdots, \mathrm{Bel}_n$ over a frame Θ is called *combinable* if it does not satisfy the conditions of Theorem 3.4. The belief function that results from combining such a collection (either iteratively or directly by (3.5)) is called its *orthogonal sum* and denoted $\mathrm{Bel}_1 \oplus \cdots \oplus \mathrm{Bel}_n$. It follows from Theorem 3.1 that the core of $\mathrm{Bel}_1 \oplus \cdots \oplus \mathrm{Bel}_n$ is equal to the intersection of the cores of the Bel_i.

As the reader who is familiar with n-dimensional geometry will have surmised, the construction of §1 can be directly generalized to describe the orthogonal combination of n belief functions. In this description, one would represent the probability mass for each belief function by a unit line segment, orthogonally combine these line segments to obtain an n-dimensional cube, identify the focus of each portion of that cube, eliminate the portions committed to $\emptyset$, and renormalize the measure of the remainder. The resulting belief function would be the same as the one given by (3.5), and in fact the renormalizing constant necessitated by the elimination of part of the cube would be the same as the renormalizing constant (3.6).

§4. The Weight of Conflict

When I explained in §1 how to combine two belief functions Bel_1 and Bel_2, I treated the renormalizing constant (3.1) as a nuisance. And it does complicate the description of Dempster's rule and the calculation of orthogonal sums. But if we examine the constant closely, we find that it has a useful intuitive role: it measures the extent of conflict between the two belief functions.

Indeed, every instance in which a rectangle in Figure 3.3 is committed to $\emptyset$ corresponds to an instance in which Bel_1 and Bel_2 commit probability to disjoint (or contradictory) subsets A_i and B_j. The greater the number of such instances, and the greater the amounts of probability that are conflictingly committed in each instance, the greater the total probability

$$\kappa = \sum_{\substack{i,j \\ A_i \cap B_j = \emptyset}} m_1(A_i)\, m_2(B_j)$$

that has to be eliminated. Since the renormalizing constant

$$K = \frac{1}{1-\kappa}$$

is increasing with κ, it can obviously serve as a measure of the extent of the conflict.

Actually, the most useful measure of the conflict between Bel_1 and Bel_2 is the quantity

$$\log K = \log \frac{1}{1-\kappa} = -\log(1-\kappa)\ ,$$

which I call the *weight of conflict between* Bel_1 *and* Bel_2 and denote $Con(Bel_1, Bel_2)$: This quantity deserves to be called a weight in part because it may take any value from zero to infinity – in contrast to K itself, which is always greater than or equal to one. If Bel_1 and Bel_2 do not conflict at all, then $\kappa = 0$ and $Con(Bel_1, Bel_2) = 0$; if Bel_1 and Bel_2 flatly contradict each other so that $Bel_1 \oplus Bel_2$ does not exist, then $\kappa = 1$ and $Con(Bel_1, Bel_2) = \infty$.

The notion of weight of conflict extends quite naturally to a collection $Bel_1, \cdots, Bel_n$ of belief functions. For as I asserted in the preceding section, the combination of these belief functions can be described in terms of an n-dimensional cube, and if κ denotes the amount of that cube that is committed to $\emptyset$ and hence must be eliminated, then the renormalizing constant (3.6) is given by

$$K = \frac{1}{1-\kappa}$$

And hence it is natural to call $\log K$ the *weight of conflict among* $Bel_1, \cdots, Bel_n$.

Like all weights, weights of conflict combine additively.

THEOREM 3.5. *Suppose* $Bel_1, \cdots, Bel_{n+1}$ *are belief functions over a frame* Θ, *and suppose* $Bel_1 \oplus \cdots \oplus Bel_n$ *exists. Then*

$$Con(Bel_1, \cdots, Bel_{n+1}) = Con(Bel_1, \cdots, Bel_n) + Con(Bel_1 \oplus \cdots \oplus Bel_n, Bel_{n+1}).$$

§5. Conditioning Belief Functions

Dempster's rule of combination permits a simple description of how the assimilation of new evidence should change our beliefs: if our initial beliefs are expressed by a belief function Bel_1 over Θ, and the new evidence alone determines a belief function Bel_2 over Θ, then after assimilating the new evidence we should have the beliefs given by $Bel_1 \oplus Bel_2$. This description avoids the doctrine that a body of evidence can always be cast in the form of a single proposition known with certainty – a doctrine long ago refuted by F. P. Ramsey and others but, as we saw in Chapter 1, still essential to the Bayesian theory. Yet it permits the assimilation of new certainties as a special case.

Suppose, indeed, that the effect of the new evidence on the frame of discernment Θ is to establish a particular subset $B \subset \Theta$ with certainty. Then Bel_2 will give a degree of belief one to the proposition corresponding to B and to every proposition implied by it:

$$Bel_2(A) = \begin{cases} 1 & \text{if } B \subset A \\ 0 & \text{if } B \not\subset A. \end{cases} \tag{3.7}$$

(The subset B is the only focal element of Bel_2, and its basic probability number is one.) Such a function Bel_2 is combinable with Bel_1 as long as $\mathrm{Bel}_1(\bar{B}) < 1$, and the orthogonal sum $\mathrm{Bel}_1 \oplus \mathrm{Bel}_2$ is very simply described.

THEOREM 3.6. *Suppose* Bel_2 *is defined by (3.7), and* Bel_1 *is another belief function over* Θ. *Then* Bel_1 *and* Bel_2 *are combinable if and only if* $\mathrm{Bel}_1(\bar{B}) < 1$. *If* Bel_1 *and* Bel_2 *are combinable, let* $\mathrm{Bel}_1(\cdot\,|B)$ *denote* $\mathrm{Bel}_1 \oplus \mathrm{Bel}_2$, *and let* P_1^* *and* $P_1^*(\cdot\,|B)$ *denote the upper probability functions for* Bel_1 *and* $\mathrm{Bel}_1 \oplus \mathrm{Bel}_2$, *respectively. Then*

$$\mathrm{Bel}_1(A|B) = \frac{\mathrm{Bel}_1(A \cup \bar{B}) - \mathrm{Bel}_1(\bar{B})}{1 - \mathrm{Bel}_1(\bar{B})}$$

and

$$P_1^*(A|B) = \frac{P_1^*(A \cap B)}{P_1^*(B)} \tag{3.8}$$

for all $A \subset \Theta$.

Since new evidence rarely occurs in the form of a certainty, these formulae are of little practical value. It is interesting though, to note the similarity of (3.8) to Bayes' rule of conditioning. I call (3.8) *Dempster's rule of conditioning.*

§6. Other Properties of Dempster's Rule

We shall become much better acquainted with Dempster's rule of combination in the next chapter, but two of its properties can be pointed out immediately.

THEOREM 3.7. *Suppose* Bel_1 *and* Bel_2 *are belief functions over* Θ. *Then*

(1) *if* Bel_1 *is vacuous, then* Bel_1 *and* Bel_2 *are combinable and* $\mathrm{Bel}_1 \oplus \mathrm{Bel}_2 = \mathrm{Bel}_2$; *and*

(2) *if* Bel_1 *is Bayesian and* Bel_1 *and* Bel_2 *are combinable, then* $\mathrm{Bel}_1 \oplus \mathrm{Bel}_2$ *is Bayesian.*

§7. Mathematical Appendix

Proof of Theorem 3.1. Since (3.2) is assumed, the function m is certainly well-defined by (3.3), and takes non-negative values. Since $m(\emptyset) = 0$ by definition, m is a basic probability assignment provided only that the $m(A)$ sum to one. And, in fact,

$$\sum_{A\subset\Theta} m(A) = m(\emptyset) + \sum_{\substack{A\subset\Theta\\ A\neq\emptyset}} m(A)$$

$$= \sum_{\substack{A\subset\Theta\\ A\neq\emptyset}} \frac{\displaystyle\sum_{\substack{i,j\\ A_i\cap A_j = A}} m_1(A_i)m_2(B_j)}{\displaystyle 1-\sum_{\substack{i,j\\ A_i\cap B_j=\emptyset}} m_1(A_i)m_2(B_j)}$$

$$= \left(\frac{1}{\displaystyle 1-\sum_{\substack{i,j\\ A_i\cap B_j=\emptyset}} m_1(A_i)m_2(B_j)}\right) \sum_{\substack{i,j\\ A_i\cap B_j\neq\emptyset}} m_1(A_i)m_2(B_j) = 1\,.$$

The core of the belief function given by m will be

$$\bigcup_{i,j} (A_i\cap B_j) = \left(\bigcup_i A_i\right)\cap\left(\bigcup_j B_j\right),$$

and $\bigcup_i A_i$ and $\bigcup_j B_j$ are the cores of Bel_1 and Bel_2, respectively. ■

Proof of Theorem 3.2.

(1) is equivalent to (2). Since the A_i and the B_j are the focal elements, $m_1(A_i)m_2(B_j) > 0$ for all i, j. And

$$\sum_{i,j} m_1(A_i)m_2(B_j) = \left(\sum_i m_1(A_i)\right)\left(\sum_j m_2(B_j)\right) = 1\,.$$

Therefore (3.2) fails if and only if $A_i \cap B_j = \emptyset$ for all i, j. If $\mathcal{C}_1$ and $\mathcal{C}_2$ denote the cores of Bel_1 and Bel_2, respectively, then

$$\mathcal{C}_1 \cap \mathcal{C}_2 = \left(\bigcup_i A_i\right) \cap \left(\bigcup_j B_j\right)$$

$$= \bigcup_{i,j} (A_i \cap B_j),$$

whence $A_i \cap B_j = \emptyset$ for all i, j if and only if $\mathcal{C}_1 \cap \mathcal{C}_2 = \emptyset$.

(3) is equivalent to (2). $Bel_1(A) = Bel_2(\bar{A}) = 1$ if and only if $\mathcal{C}_1 \subset A$ and $\mathcal{C}_2 \subset \bar{A}$. So there exists $A \subset \Theta$ such that $Bel_1(A) = Bel_2(\bar{A}) = 1$ if and only if $\mathcal{C}_1 \cap \mathcal{C}_2 = \emptyset$.

(4) is equivalent to (2). A belief function gives a positive commonality number to any singleton contained in its core, and a zero commonality number to any subset not contained in its core. So if $\mathcal{C}_1 \cap \mathcal{C}_2 \neq \emptyset$, then $Q_1(\{\theta\})Q_2(\{\theta\}) > 0$ for any $\theta \in \mathcal{C}_1 \cap \mathcal{C}_2$. And if $\mathcal{C}_1 \cap \mathcal{C}_2 = \emptyset$, then every non-empty $A \subset \Theta$, since it will not be a subset of both $\mathcal{C}_1$ and $\mathcal{C}_2$, will satisfy either $Q_1(A) = 0$ or $Q_2(A) = 0$ and hence will satisfy $Q_1(A)Q_2(A) = 0$. ∎

Proof of Theorem 3.3. When the quantity (3.1) is denoted by K, (3.3) yields

$$Q(A) = \sum_{\substack{B \subset \Theta \\ A \subset B}} m(B)$$

$$= K \sum_{\substack{B \subset \Theta \\ A \subset B}} \sum_{\substack{i,j \\ A_i \cap B_j = B}} m_1(A_i)m_2(B_j)$$

$$= K \sum_{\substack{i,j \\ A \subset A_i \cap B_j}} m_1(A_i)m_2(B_j)$$

$$= K \sum_{\substack{i,j \\ A \subset A_i \\ A \subset B_j}} m_1(A_i) m_2(B_j)$$

$$= K \left(\sum_{\substack{i \\ A \subset A_i}} m_1(A_i) \right) \left(\sum_{\substack{j \\ A \subset B_j}} m_2(B_j) \right)$$

$$= K \left(\sum_{\substack{B \subset \Theta \\ A \subset B}} m_1(B) \right) \left(\sum_{\substack{B \subset \Theta \\ A \subset B}} m_2(B) \right)$$

$$= K Q_1(A) Q_2(A)$$

for all non-empty $A \subset \Theta$. ■

Proof of Theorem 3.4.

(1) is equivalent to (2). Again, a belief function gives a positive commonality number to any singleton contained in its core, and a zero commonality number to any subset not contained in its core. So if $\mathcal{C}_1 \cap \cdots \cap \mathcal{C}_n \neq \emptyset$, then $Q_1(\{\theta\}) \cdots Q_n(\{\theta\}) > 0$ for any $\theta \in \mathcal{C}_1 \cap \cdots \cap \mathcal{C}_n$. And if $\mathcal{C}_1 \cap \cdots \cap \mathcal{C}_n = \emptyset$, then every non-empty $A \subset \Theta$, since it will not be a subset of all the $\mathcal{C}_i$, will satisfy $Q_i(A) = 0$ for some i and hence will satisfy $Q_1(A) \cdots Q_n(A) = 0$.

(1) is equivalent to (3). If (1) holds, then the $\mathcal{C}_i$ satisfy the conditions that (3) requires of its A_i. If (3) holds, then $\mathcal{C}_i \subset A_i$ for each i, and $\mathcal{C}_1 \cap \cdots \cap \mathcal{C}_n$ is empty because it is contained in $A_1 \cap \cdots \cap A_n$.

(3) is equivalent to (4). If (3) holds, then one may obtain (4) by setting $A = A_1$ and $B = A_2 \cap \cdots \cap A_n$. If (4) holds, then one may obtain (3) by setting A_i equal to the intersection of those of the $E_1, \cdots, E_r, F_1, \cdots, F_s$ that are awarded degree of belief one by Bel_i, provided some are, and equal to Θ if none are.

(2) is equivalent to (5) and to (6). This is obvious from Theorems 3.1 and 3.2. ■

Proof of Theorem 3.5. Let $Q_1,\cdots,Q_n$, Q, and Q' denote the commonality functions for $Bel_1,\cdots,Bel_n$, $Bel_1\oplus\cdots\oplus Bel_n$, and $Bel_1\oplus\cdots\oplus Bel_{n+1}$, respectively. Notice that by definition,

$$\mathrm{Con}(Bel_1,\cdots,Bel_n) = \log K ,$$

$$\mathrm{Con}(Bel_1,\cdots,Bel_{n+1}) = \log K' ,$$

and

$$\mathrm{Con}(Bel_1\oplus\cdots\oplus Bel_n, Bel_{n+1}) = \log K'' ,$$

where K is the renormalizing constant for combining $Bel_1,\cdots,Bel_n$, K' is the renormalizing constant for combining $Bel_1,\cdots,Bel_{n+1}$, and K'' is the renormalizing constant for combining $Bel_1\oplus\cdots\oplus Bel_n$ with Bel_{n+1}. Hence

$$Q(A) = KQ_1(A)\cdots Q_n(A) ,$$

$$Q'(A) = K'Q_1(A)\cdots Q_n(A)Q_{n+1}(A) ,$$

and

$$Q'(A) = K''Q(A)Q_{n+1}(A)$$

for all non-empty $A \subset \Theta$. Choosing a non-empty $A \subset \Theta$ for which $Q'(A)>0$ and substituting the first two equations in the third, one finds that

$$K' = KK'' ,$$

or

$$\log K' = \log K + \log K'' . \blacksquare$$

Proof of Theorem 3.6. $Bel_1(\bar{B}) < 1$ if and only if B overlaps the core of Bel_1, and since B is the core of Bel_2, this is indeed equivalent to Bel_1 being combinable with Bel_2.

Denote the basic probability assignments of Bel_1, Bel_2 and $Bel_1\oplus Bel_2$ by m_1, m_2 and m. Since B is the only focal element of Bel_2, and $m_2(B) = 1$, (3.3) yields

$$m(A) = \frac{\sum\limits_{\substack{i \\ A_i \cap B = A}} m_1(A_i)}{1 - \sum\limits_{\substack{i \\ A_i \cap B = \emptyset}} m_1(A_i)} = \frac{\sum\limits_{\substack{C \\ B \cap C = A}} m_1(C)}{1 - Bel_1(\bar{B})}$$

and

$$Bel_1(A|B) = \sum_{D \subset A} m(D) = \frac{\sum\limits_{\substack{D \\ \emptyset \neq D \subset A}} \sum\limits_{\substack{C \\ B \cap C = D}} m_1(C)}{1 - Bel_1(\bar{B})}$$

$$= \frac{\sum\limits_{\substack{C \\ \emptyset \neq B \cap C \subset A}} m_1(C)}{1 - Bel_1(\bar{B})}$$

$$= \frac{\sum\limits_{\substack{C \subset A \cup \bar{B} \\ C \not\subset \bar{B}}} m_1(C)}{1 - Bel_1(\bar{B})}$$

$$= \frac{Bel_1(A \cup \bar{B}) - Bel_1(\bar{B})}{1 - Bel_1(\bar{B})} .$$

Hence

$$P_1^*(A|B) = 1 - Bel_1(\bar{A}|B)$$

$$= \frac{1 - Bel_1(\bar{B}) - Bel_1(\bar{A} \cup \bar{B}) + Bel_1(\bar{B})}{1 - Bel_1(\bar{B})}$$

$$= \frac{1 - Bel_1(\overline{A \cap B})}{1 - Bel_1(\bar{B})}$$

$$= \frac{P_1^*(A \cap B)}{P_1^*(B)} . \blacksquare$$

Proof of Theorem 3.7.

(i) Denote the commonality functions for Bel_1 and Bel_2 by Q_1 and Q_2, and their cores by $\mathcal{C}_1$ and $\mathcal{C}_2$. If Bel_1 is vacuous, then $\mathcal{C}_1 = \Theta$ and $Q_1(A) = 1$ for all $A \subset \Theta$. It follows by Theorem 3.2 that $\mathrm{Bel}_1 \oplus \mathrm{Bel}_2$ exists, and by Theorem 3.3 together with (2.6) that it has the same commonality function as Bel_2.

(ii) By Theorem 2.8, a belief function is Bayesian if and only if its commonality numbers for non-singletons are all zero. And by Theorem 3.3, $\mathrm{Bel}_1 \oplus \mathrm{Bel}_2$ will have zero commonality numbers for non-singletons if Bel_1 does. ∎

CHAPTER 4. SIMPLE AND SEPARABLE SUPPORT FUNCTIONS

> Many probabilities concurring preuaile much.
>
> THOMAS GRANGER (b. 1578)

This chapter performs three tasks. First, it justifies Dempster's rule by exhibiting the sensible and intuitive results of that rule in the simplest cases. Second, it introduces the notion of weight of evidence. And third, it introduces two classes of belief functions that are appropriate for the representation of evidence: the *simple support functions* and the larger class of *separable support functions*.

Simple support functions are belief functions based on homogeneous evidence – evidence whose impact on one's frame of discernment is precisely to support a given subset of it. Separable support functions include both simple support functions and orthogonal sums of simple support functions.

As the vocabulary suggests, we will later encounter a yet larger class of belief functions called *support functions*. Schematically:

$$\left\{\begin{matrix}\text{simple}\\ \text{support}\\ \text{functions}\end{matrix}\right\} \subset \left\{\begin{matrix}\text{separable}\\ \text{support}\\ \text{functions}\end{matrix}\right\} \subset \left\{\begin{matrix}\text{support}\\ \text{functions}\end{matrix}\right\} \subset \left\{\begin{matrix}\text{belief}\\ \text{functions}\end{matrix}\right\}.$$

As we will see in Chapter 7, the class of support functions includes all those belief functions that can be obtained by coarsening the frame of a separable support function. Those belief functions that are not support functions are discussed in Chapter 9.

§1. Simple Support Functions

The effect of a body of evidence on a frame of discernment Θ is usually to support many of the propositions discerned by Θ, but to varying degrees. But the simplest situation is undoubtedly that where the evidence points precisely and unambiguously to a single non-empty subset A of Θ. In that situation, we can say that the effect of the evidence is limited to providing a certain degree of support for A. By virtue of supporting A to a certain degree, such evidence will, of course, support any proposition implied by A to an equal degree. But it will provide no support at all for any of the other propositions discerned by Θ.

The *degrees of support* provided by such evidence are thus very easy to specify. If s is the degree of support for A, where $0 \leq s \leq 1$, then the degree of support for $B \subset \Theta$ is given by

$$S(B) = \begin{cases} 0 & \text{if} \quad B \text{ does not contain } A \\ s & \text{if} \quad B \text{ contains } A \text{ but } B \neq \Theta \\ 1 & \text{if} \quad B = \Theta . \end{cases}$$

The function $S : 2^{\Theta} \to [0,1]$ thus defined is called a *simple support function* focused on A.

It is easily seen that a simple support function is a belief function. Indeed, if S is a simple support function focused on A, then S is the belief function with basic probability numbers $m(A) = S(A)$, $m(\Theta) = 1 - S(A)$, and $m(B) = 0$ for all other $B \subset \Theta$.

§2. Bernoulli's Rule of Combination

Suppose one body of evidence has the precise effect of supporting $A \subset \Theta$ to the degree s_1, while another entirely separate body of evidence has the precise effect of supporting A to the degree s_2. Then to what degree do the two bodies of evidence together support A? We require, evidently, a rule for combining two simple support functions S_1 and S_2, both of which are focused on A.

The need for such a rule of combination was recognized by James Bernoulli, one of the most prominent early students of probability.* His solution, translated into our terminology, is to set the degree of support for A, based on the total evidence, equal to

$$\begin{aligned} 1-(1-s_1)(1-s_2) &= s_1 + s_2(1-s_1) \\ &= s_2 + s_1(1-s_2) . \end{aligned}$$

This solution has the virtue, at least, of assuring that A's degree of support on the total evidence is greater than its degree of support on either of the separate bodies of evidence. Indeed, the deficit from unity of the degree of support provided by one of the bodies of evidence is reduced by a proportion equal to the degree of support provided by the other body of evidence.

Bernoulli's rule is actually a special case of Dempster's rule of combination, and we can verify this by carrying out the construction of §1 of Chapter 3. Our simple support function S_1 has the basic probability numbers $m_1(A) = s_1$ and $m_1(\Theta) = 1 - s_1$, while our simple support function S_2 has the basic probability numbers $m_2(A) = s_2$ and $m_2(\Theta) = 1 - s_2$; hence we obtain the picture in Figure 4.1. Only the upper right-hand rectangle, of measure $(1-s_1)(1-s_2)$, fails to be committed to A. So

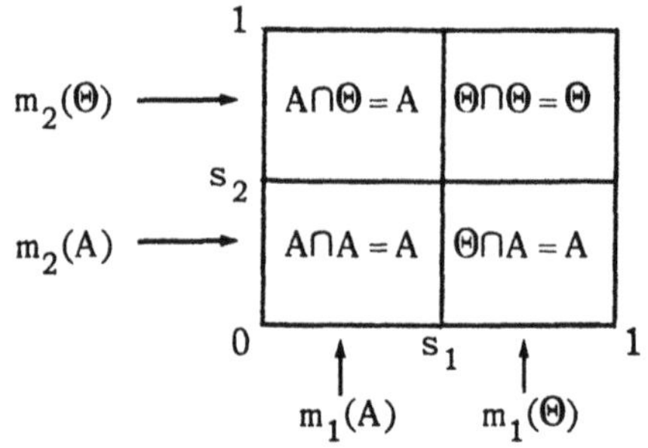

Figure 4.1.

*Bernoulli discussed the problem of combining evidence in Chapter IV of Part IV of his famous *Ars Conjectandi*. His proposals were not all entirely satisfactory, but they inspired the more successful efforts of J. H. Lambert.

the orthogonal sum $S = S_1 \oplus S_2$ has basic probability numbers $m(A) = 1 - (1-s_1)(1-s_2)$ and $m(\Theta) = (1-s_1)(1-s_2)$; S is a simple support function focused on A with $S(A) = 1 - (1-s_1)(1-s_2)$.

§3. The Weight of Evidence

Epistemic probabilities, or degrees of support, have often been associated with *weights of evidence.* It would seem that the degrees of support for the various propositions discerned by a frame Θ ought to be determined by the weights of the items of evidence attesting to those various propositions. And in the case of the evidence underlying a simple support function, this determination ought to be fairly simple. Indeed, if the evidence points precisely and unambiguously to a single subset $A \subset \Theta$, then the degree of support for A ought to be completely determined by the weight of that evidence; it ought to be a function of that weight.

Let us investigate the mathematical form of this function. While the weight w of the evidence pointing to A can take any non-negative value, the degree of support s for A must be between zero and one. Hence we require a function $g : [0, \infty] \to [0, 1]$ such that $s = g(w)$. What do we know about this function?

The function g is very nearly determined by Bernoulli's rule of combination together with the intuitive idea that weights combine additively. Indeed, if we denote by w_1 and w_2 the weights of evidence underlying the simple support functions S_1 and S_2 of the preceding section, then the weight of evidence underlying their orthogonal sum will be $w_1 + w_2$. Hence

$$g(w_1) = s_1 ,$$

$$g(w_2) = s_2 ,$$

and

$$g(w_1 + w_2) = 1 - (1-s_1)(1-s_2) .$$

And if we set $f(w) = 1 - g(w)$, then these relations become

$$f(w_1) = 1 - s_1 ,$$

$$f(w_2) = 1 - s_2 ,$$

and

$$f(w_1 + w_2) = (1 - s_1)(1 - s_2) ,$$

whence

$$f(w_1 + w_2) = f(w_1)\ f(w_2) .$$

Hence f must be exponential;*

$$f(w) = e^{cw}$$

and

$$g(w) = 1 - e^{cw} ,$$

where c is a constant.

In order for f to map $[0, \infty]$ onto $[0, 1]$, the constant c must be negative. But otherwise it is quite at our disposal, for the unit of measurement for the weights of evidence is arbitrary, or more accurately, figmentary. Setting $c = -1$ for convenience, we thus obtain

$$s = 1 - e^{-w}$$

as the relation between the weight of evidence precisely and unambiguously supporting A and the degree of support for A in the corresponding simple support function. Solving this relation for w gives

$$w = -\log(1 - s)$$

as the weight of evidence that is needed in order to produce the degree of support s.

Notice that the weight of evidence can take any non-negative value, including zero and infinity. Evidence of zero weight produces a degree of support of zero, while evidence of infinite weight produces a degree of support of one, or certainty.

---

*Assuming continuity. See, for example, p. 183 of Courant's calculus text.

§4. Heterogeneous Evidence

So far in this chapter, we have studied only homogeneous evidence – evidence that points unambiguously to a single subset. We now turn to the problem of combining bodies of evidence that point to different subsets. In this section, we consider the combination of two simple support functions, one focused on A, and one focused on B, where $A \cap B \neq \emptyset$. In the next section, we consider the case where $A \cap B = \emptyset$.

The essential feature of combining evidence that points towards a proposition A with evidence that points towards a different but compatible proposition B is that the total evidence provides support not only for A and B separately, but also for the conjunction $A \cap B$. This intuitive fact should be captured by a rule of combination.

Suppose $A \cap B \neq \emptyset$, and we wish to combine S_1 and S_2, where S_1 is a simple support function focused on A, with $S_1(A) = s_1$, and S_2 is a simple support function focused on B, with $S_2(B) = s_2$. Then to what extent, exactly, does the combined evidence provide positive support for $A \cap B$?

Dempster's rule provides a reasonable answer: $A \cap B$ is supported to the extent $s_1 s_2$. Indeed, as we see from Figure 4.2, the orthogonal sum $S = S_1 \oplus S_2$ has basic probability numbers $m(A \cap B) = s_1 s_2$, $m(A) = s_1(1-s_2)$, $m(B) = s_2(1-s_1)$, and $m(\Theta) = (1-s_1)(1-s_2)$.

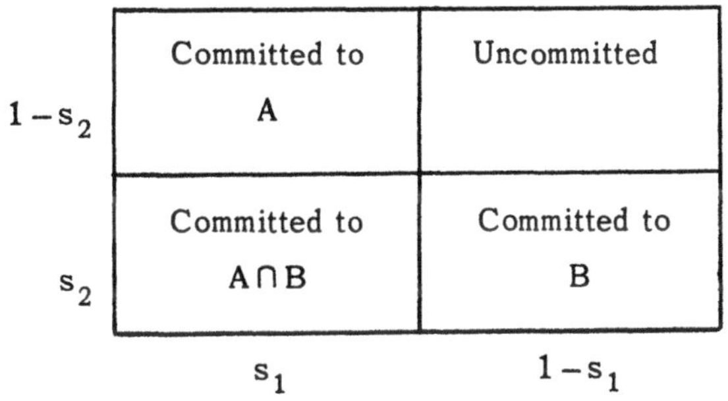

Figure 4.2.

Hence

$$S(C) = \begin{cases} 0 & \text{if } C \text{ does not contain } A \cap B \\ s_1 s_2 & \text{if } C \text{ contains } A \cap B \text{ but neither } A \text{ nor } B \\ s_1 & \text{if } C \text{ contains } A \text{ but not } B \\ s_2 & \text{if } C \text{ contains } B \text{ but not } A \\ 1-(1-s_1)(1-s_2) & \text{if } C \text{ contains both } A \text{ and } B \text{ but } C \neq \Theta \\ 1 & \text{if } C = \Theta \end{cases}$$

for all $C \subset \Theta$.

Let us apply this rule to a simple example.

EXAMPLE 4.1. *The Burglary of the Sweetshop.* Sherlock Holmes is investigating the burglary of a sweetshop. By examining the opened safe he has been able to conclude, with a high degree of certainty, that the thief was left-handed. With this as his only evidence, he will not, of course, assign any more than a miniscule degree of support to the guilt of any particular person, even a left-handed one. In particular, he will not assign any substantial degree of support to the guilt of the single one of the shop's clerks who is left-handed. But now suppose we permit Mr. Holmes a second bit of evidence, whence he is able to conclude, again with a high degree of certainty, that the theft was an "inside job." Then the evidence against left-handers will devolve, with most of its force, upon the left-handed clerk; the proposition that he is guilty will acquire a high degree of support.

This tale can be represented by using a frame that contains four possibilities, corresponding to whether the thief was left-handed or right-handed, an insider or an outsider. That is to say, we may set

$$\Theta = \{LI, LO, RI, RO\}$$

where LI means that the thief was a left-handed insider, etc. The first body of evidence, which indicates a left-handed thief, induces a simple support function S_1 focused on $A = \{LI, LO\}$, while the second body of evidence, which indicates an insider, induces a simple support function S_2 focused on $B = \{LI, RI\}$. Combining S_1 and S_2 by Dempster's rule produces a positive degree of support for $\{LI\} = A \cap B$. In fact, $S(\{LI\}) = S_1(A)S_2(B)$. If both $S_1(A)$ and $S_2(B)$ are very high, then $S(\{LI\})$ will be very high. And if $S_2(B) = 1$ – if Holmes can be completely certain that the theft was an "inside job" – then $S(\{LI\}) = S_1(A)$; the total weight of evidence against left-handers will devolve upon the left-handed clerk. ■

So far I have tacitly assumed that neither A nor B contains the other. If A is a subset of B, say, then the evidence is not really so heterogeneous, and the expression for the orthogonal sum simplifies to

$$S(C) = \begin{cases} 0 & \text{if } C \text{ does not contain } A \\ s_1 & \text{if } C \text{ contains } A \text{ but not } B \\ 1-(1-s_1)(1-s_2) & \text{if } C \text{ contains } B \text{ but } C \neq \Theta \\ 1 & \text{if } C = \Theta. \end{cases}$$

The following example illustrates this simpler situation.

EXAMPLE 4.2. *The Cabbage Seed.* A plant has just sprouted in a pot. There is some evidence that the plant is a cabbage; to wit: I planted a cabbage seed in the pot. Having modest confidence in my gardening skill, I find that this evidence provides a degree of support of $\frac{1}{2}$, say, for the plant being a cabbage. Upon investigating the plant, I discover it has two leaves; making due allowance for the possibility that either I or the plant made a mistake, this provides a degree of support of $\frac{9}{10}$, say, for the plant being a dicotyledon. What is the effect of combining the two items of evidence?

To formalize the problem, denote the set of the various species of plants by Θ, denote the set of dicotyledons by B, and let A = {cabbage}. Then $A \subset B \subset \Theta$. And our problem is that of combining a simple support function S_1 focused on A with $S_1(A) = .50$ and a simple support function S_2 focused on B with $S_2(B) = .90$. According to the preceding formula, the orthogonal sum $S = S_1 \oplus S_2$ satisfies $S(A) = .50$ and $S(B) = .95$; the evidence for cabbage enhances the support for dicotyledons, but the evidence for dicotyledons does not enhance the support for cabbage. ■

§5. Conflicting Evidence

While the combination of evidence pointing to A with evidence pointing to B is quite simple when $A \cap B \neq \emptyset$, it is more complicated when $A \cap B = \emptyset$. For then the two bodies of evidence are not just heterogeneous; they are conflicting, and the effect of each is diminished by the other.

Figure 4.3 shows what happens when the scheme of the preceding section is applied to this case: the lower-left rectangle is committed to $\emptyset$. To apply Dempster's rule, we must therefore eliminate this rectangle and inflate the measures of the remaining rectangles by the factor

$$\frac{1}{1 - s_1 s_2} .$$

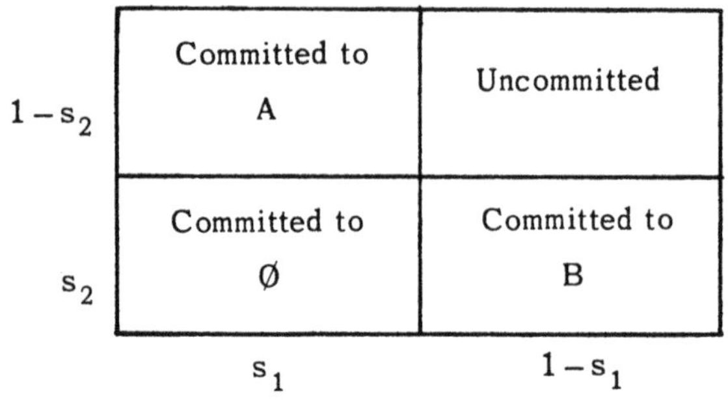

Figure 4.3.

Hence the orthogonal sum S of the simple support function S_1 (focused on A with $S_1(A) = s_1$) and the simple support function S_2 (focused on B with $S_2(B) = s_2$) is given by the basic probability numbers $m(A) = \frac{s_1(1-s_2)}{1-s_1s_2}$, $m(B) = \frac{s_2(1-s_1)}{1-s_1s_2}$ and $m(\Theta) = \frac{(1-s_1)(1-s_2)}{1-s_1s_2}$;

$$S(C) = \begin{cases} 0 & \text{if } C \text{ contains neither } A \text{ nor } B \\ \dfrac{s_1(1-s_2)}{1-s_1s_2} & \text{if } C \text{ contains } A \text{ but not } B \\ \dfrac{s_2(1-s_1)}{1-s_1s_2} & \text{if } C \text{ contains } B \text{ but not } A \\ \dfrac{s_1(1-s_2)+s_2(1-s_1)}{1-s_1s_2} & \text{if } C \text{ contains } A \text{ and } B \text{ but } C \neq \Theta \\ 1 & \text{if } C = \Theta \end{cases}$$

for all $C \subset \Theta$. (Notice that this orthogonal sum exists if and only if not both s_1 and s_2 equal one; we cannot combine certainty in A with certainty in B.)

This result seems reasonable, especially because it accords with the apprehension that each of the two bodies of evidence ought to reduce the effect of the other. Notice, for example, that the introduction of the second body of evidence reduces the degree of support for A from s_1 to

$$\frac{s_1(1-s_2)}{1-s_1s_2} = s_1 \frac{1-s_2}{1-s_1s_2} .$$

The factor $\frac{1-s_2}{1-s_1s_2}$ is decreasing in s_2; the greater the support for B, the more the support for A is eroded.

Further insight can be gained by relating the degrees of support $S(A)$ and $S(B)$ to the weights of evidence. Denoting the weights of evidence pointing to A and B by $w(A)$ and $w(B)$, respectively, we can use the relations

$$s_1 = 1 - e^{-w(A)}$$

and

$$s_2 = 1 - e^{-w(B)}$$

to obtain

$$S(A) = \frac{s_1(1-s_2)}{1 - s_1 s_2} = \frac{e^{w(A)} - 1}{e^{w(A)} + e^{w(B)} - 1}$$

and

$$S(B) = \frac{s_2(1-s_1)}{1 - s_1 s_2} = \frac{e^{w(B)} - 1}{e^{w(A)} + e^{w(B)} - 1} .$$

These relations can be solved for $w(A)$ and $w(B)$;

$$w(A) = \log \frac{1 - S(B)}{1 - S(A) - S(B)}$$

and

$$w(B) = \log \frac{1 - S(A)}{1 - S(A) - S(B)}$$

are the weights of evidence required to produce degrees of support $S(A)$ and $S(B)$ for two contradictory propositions A and B.

EXAMPLE 4.3. *The Alibi.* A criminal defendant has an alibi: a close friend swears that the defendant was visiting his apartment at the time of the crime. The friend has a good reputation, so his testimony carries some weight in spite of his close friendship with the defendant; let us suppose that standing alone it would provide a degree of support of $\frac{1}{10}$ for the defendant's innocence. But ranged on the other side is a strong body of circumstantial evidence attesting to the defendant's guilt; standing alone it would provide a degree of support of $\frac{9}{10}$ for his guilt. What degrees of support does the combined evidence provide for the defendant's guilt and innocence?

To formalize the example, we set $\Theta = \{G, I\}$, where G stands for guilt and I for innocence, we denote by S_1 the simple support function focused on $\{I\}$ with $S_1(\{I\}) = \frac{1}{10}$ and we denote by S_2 the simple support function focused on $\{G\}$ with $S_2(\{G\}) = \frac{9}{10}$. The orthogonal sum $S = S_1 \oplus S_2$ then yields

$$S(\{I\}) = \frac{1}{91},$$

and

$$S(\{G\}) = \frac{81}{91};$$

the effect of the friend's testimony has mildly eroded the force of the circumstantial evidence, yet has been strongly eroded by it. ■

EXAMPLE 4.4. *A Biased Coin.* A certain coin is known to be biased either to heads or to tails. More precisely, it is known that when the coin is tossed in a certain way, it is governed either by the chance density q_{θ_1} or by the chance density q_{θ_2}; q_{θ_1} assigns a chance of .9 to heads and a chance of .1 to tails, while q_{θ_2} assigns a chance of .3 to heads and a chance of .7 to tails. (The reader is asked to leave aside the question of the origin of such knowledge.) Suppose we toss the coin and obtain heads. Since q_{θ_1} assigns a greater chance to heads than q_{θ_2}, this outcome must constitute evidence in favor of q_{θ_1}. A second, physically independent toss resulting in tails will similarly provide evidence in favor of q_{θ_2}. What degrees of support are provided by these two items of evidence singly, and how should the two items be combined?

It is natural, when only two chance densities are considered, to assess the weight of evidence by taking the logarithm of the ratio of the chance assigned the actual outcome by the one density to the chance assigned it by the other.* Thus the outcome of

*I. J. Good, an advocate of the Bayesian theory, has used the term "weight of evidence" to name this logarithm. For an explanation of its role in the

heads on the first toss provides a weight of evidence of

$$\log \frac{.9}{.3} = \log 3$$

in favor of q_{θ_1}, while the outcome of tails on the second toss provides a weight of evidence of

$$\log \frac{.7}{.1} = \log 7$$

in favor of q_{θ_2}. Using these two weights, we may calculate degrees of support based on either toss separately or on both together. Setting $\Theta = \{\theta_1, \theta_2\}$, we find that the first toss produces a simple support function S_1 focused on $\{\theta_1\}$ with $S_1(\{\theta_1\}) = \frac{2}{3}$, while the second toss produces a simple support function S_2 focused on $\{\theta_2\}$ with $S_2(\{\theta_2\}) = \frac{6}{7}$. Setting $S = S_1 \oplus S_2$, we obtain $S(\{\theta_1\}) = \frac{2}{9}$ and $S(\{\theta_2\}) = \frac{2}{3}$. ■

§6. Separable Support Functions

In the examples of this chapter, Dempster's rule has been applied only to pairs of simple support functions. These examples have been sufficient, though, to illustrate all the essential features of the rule, and they have shown those features to be reasonable and useful. Hence we can feel confident in applying the rule not only to pairs, but also to larger collections of simple support functions.

Bayesian theory, see p. 63 of Good's *Probability and the Weighing of Evidence*. Our adoption of the convention that the weight of evidence is equal to this logarithm is based on its intuitive appropriateness and does not constitute an appeal to the Bayesian theory.

Since we are accustomed to thinking about chances, this convention may help establish a standard for weighing other types of evidence. Indeed, when we are asked to identify the unit of measurement for weights of evidence, we may explain that an observation twice as likely under one chance density as under a second chance density provides a weight of evidence of log 2 in favor of the first density, that an observation ten times as likely provides a weight of log 10, etc.

I call a belief function a *separable support function* if it is a simple support function or is equal to the orthogonal sum of two or more simple support functions. The separable support functions do not include all the belief functions that are appropriate for the representation of evidence. But a separable support function is appropriate whenever the evidence can be decomposed into components that are homogeneous with respect to one's frame of discernment.

CHAPTER 5. THE WEIGHTS OF EVIDENCE

> The magnitude of the probability of an argument ··· depends upon a balance between what may be termed the favourable and the unfavourable evidence.
>
> JOHN MAYNARD KEYNES (1883-1946)

We have learned to translate weights of evidence into simple support functions, and we have learned to combine those simple support functions into a separable support function. This chapter focuses on the net result: a rule for converting a collection of weights of evidence into a collection of degrees of support.

Given a separable support function, one would like to be able to recover the simple support functions that combined to form it and hence the weights of evidence underlying it. Unfortunately, this ambition cannot quite be satisfied, for the representation of a separable support function as the orthogonal sum of simple support functions is never quite unique. In the case where the core of the separable support function is the whole frame Θ, this non-uniqueness is not too troublesome, for the total weight of evidence focused on each subset is the same no matter what representation is used. But in the case where the core is a proper subset of the whole frame, not even these total weights are quite unique. In §3 below we resolve this difficulty by replacing the original weights with the *assessment of evidence*, a function $w : 2^{\Theta} \to [0, \infty]$ such that $w(A)$ is the total weight of evidence "effectively" focused on A.

In §4, we use the possibility of decomposing a separable support function into simple support functions to define the *weight of internal conflict* for the separable support function. In §5, we shift our attention from the total weights of evidence focused on the various subsets to the total weights of evidence impinging on the various subsets, and we use these weights together with the weight of internal conflict to formulate a more intuitively accessible version of the rule for converting weights of evidence into degrees of support. And in §6, we use the impingement function and the weight of internal conflict to formulate some conjectures that will come into play in Chapter 8.

§1. Decomposing Separable Support Functions

A separable support function is, by definition, a belief function that can be decomposed into simple support functions. That is to say, every separable support function S can be written in the form

$$S = S_1 \oplus \cdots \oplus S_n ,$$

where $n \geq 1$ and each S_i is a simple support function. But this decomposition is never unique.

Let us list some of the ways in which one may vary the decomposition of a separable support function.

THEOREM 5.1. *Suppose* S *is a separable support function over* Θ, *and* $S = S_1 \oplus \cdots \oplus S_n$, *where each of the* S_i *is a simple support function. Let* $\mathcal{C}$ *denote the core of* S, *and let* $A_1, \cdots, A_n$ *denote the foci of* $S_1, \cdots, S_n$.

(1) *Suppose* S_{n+1} *is the vacuous belief function. Then* S_{n+1} *is a simple support function, and* $S = S_1 \oplus \cdots \oplus S_{n+1}$.

(2) *If* $A_1 = A_2$, *then* $S_1 \oplus S_2$ *is a simple support function, and* $S = (S_1 \oplus S_2) \oplus S_3 \oplus \cdots \oplus S_n$.

(3) *Suppose* S_{n+1} *is a simple support function with focus* A_{n+1} *such that* $S_{n+1}(A_{n+1}) < 1$ *and* $A_{n+1} \subset \overline{\mathcal{C}}$. *Then* $S = S_1 \oplus \cdots \oplus S_{n+1}$.

(4) *Suppose all the foci* A_i *have non-empty intersections with* $\mathcal{C}$. *For each* i, *let* S'_i *denote the simple support function focused on* $A_i \cap \mathcal{C}$ *with* $S'_i(A_i \cap \mathcal{C}) = S_i(A_i)$. *Then* $S = S'_1 \oplus \cdots \oplus S'_n$.

Assertions (3) and (4) of this theorem reflect the certainty accorded to $\mathcal{C}$. This certainty overwhelms any contrary but uncertain evidence underlying a simple support function S_{n+1}, and it shifts any support for a subset A_i to the smaller subset $A_i \cap \mathcal{C}$.

As Theorem 5.1 suggests, the decomposition of a separable support function into simple support functions will be unique if one requires the simple support functions to be non-vacuous and to have distinct foci all contained in the core $\mathcal{C}$:

THEOREM 5.2. *Suppose* S *is a non-vacuous separable support function with core* $\mathcal{C}$. *Then there exists a unique collection* $S_1, \cdots, S_n$ *of non-vacuous simple support functions satisfying the following conditions:*

(1) $n \geq 1$.

(2) $S = S_1$ *if* $n = 1$, *and* $S = S_1 \oplus \cdots \oplus S_n$ *if* $n \geq 1$.

(3) *The focus of each* S_i *is contained in* $\mathcal{C}$.

(4) *If* $i \neq j$, *then the focus of* S_i *is not equal to the focus of* S_j.

The unique decomposition $S_1, \cdots, S_n$ is called the *canonical decomposition* of S. Theorem 5.1 might prompt one to guess that the canonical decomposition can be obtained from an arbitrary decomposition by omitting all the simple support functions that are vacuous or have foci not intersecting the core, reducing the other foci to their intersections with the core, and combining any of the resulting simple support functions that have common foci. The correctness of this guess will become obvious in §3 below, after we have recast our discussion in terms of the weights of evidence.

§2. Combining Weights of Evidence

Being an orthogonal sum of simple support functions, a separable support function may be described directly in terms of the weights of evidence underlying those simple support functions.

Suppose, indeed, that the separable support function S is the orthogonal sum of simple support functions $S_1, \cdots, S_n$ focused on the proper non-empty subsets $A_1, \cdots, A_n$ of Θ and based on the weights of evidence $w_1, \cdots, w_n$. Then since S_i assigns the basic probability number $1 - e^{-w_i}$ to A_i and the basic probability number e^{-w_i} to Θ, its commonality function Q_i is given by

$$Q_i(A) = \begin{cases} 1 & \text{if } A \subset A_i \\ e^{-w_i} & \text{if } A \not\subset A_i . \end{cases}$$

And according to (3.5), the commonality function Q for S is therefore given by

$$\begin{aligned} Q(A) &= K \prod_{i=1}^{n} Q_i(A) \\ &= K \prod_{\substack{i \\ A \not\subset A_i}} e^{-w_i} \\ &= K \exp\Bigg(-\sum_{\substack{i \\ A \not\subset A_i}} w_i\Bigg) \end{aligned} \tag{5.1}$$

for all non-empty $A \subset \Theta$, where

$$K = e^{\operatorname{Con}(S_1, \cdots, S_n)} = \Bigg(\sum_{\substack{B \subset \Theta \\ B \neq \emptyset}} (-1)^{|B|+1} \exp\Bigg(-\sum_{\substack{i \\ B \not\subset A_i}} w_i\Bigg)\Bigg)^{-1} . \tag{5.2}$$

As we see from the following theorem, we can also judge directly from the weights of evidence whether a collection of simple support functions is combinable and what the core of their orthogonal sum will be.

THEOREM 5.3. *Suppose* $S_1, \cdots, S_n$ *are simple support functions focused on the proper non-empty subsets* $A_1, \cdots, A_n$ *of* Θ *and based on the weights of evidence* $w_1, \cdots, w_n$. *Then* $S_1 \oplus \cdots \oplus S_n$ *exists if and only if*

$$\mathcal{C} \equiv \bigcap_{\substack{i \\ w_i = \infty}} A_i \neq \emptyset \; . \tag{5.3}$$

(This formula is to be understood in such a way that $\mathcal{C} = \Theta$ *when all the weights* w_i *are finite.) Suppose* $S_1 \oplus \cdots \oplus S_n$ *does exist. Then its core is equal to* $\mathcal{C}$*; its commonality function* Q *is given by (5.1); and* $Q(A) > 0$ *if and only if* $A \subset \mathcal{C}$.

Notice the last clause of this theorem: a separable support function awards a positive commonality number to its core. As we saw in Example 2.1, not all belief functions do this.

The representation of the commonality function Q as a function of the weights of evidence helps us see more clearly why and how the decomposition of a separable support function is not unique. In particular, it provides us with an easy way of proving Theorem 5.1:

Proof of Theorem 5.1.

(i) It is evident from (5.1) and (5.2) that one may vary the w_i without changing Q if one does not vary the sums

$$\sum_{\substack{i \\ B = A_i}} w_i$$

for the various $B \subset \Theta$. And this fact establishes assertions (1) and (2) of Theorem 5.1.

(ii) The addition of a finite weight w_{n+1} focused on a subset $A_{n+1} \subset \mathcal{C}$ will multiply the quantity

$$\exp\left(- \sum_{\substack{i \\ B \not\subset A_i}} w_i\right) \tag{5.4}$$

by $e^{-w_{n+1}}$ for all $B \not\subset A_{n+1}$. In fact, one may say that it will multiply (5.4) by $e^{-w_{n+1}}$ for all non-empty $B \subset \Theta$, for (5.4) is zero in the case of those B that are contained in $A_{n+1} \subset \overline{\mathcal{C}}$. Hence the addition of w_{n+1} multiplies the factor K in (5.1) by $e^{w_{n+1}}$, multiplies the other factor by $e^{-w_{n+1}}$, and leaves the product Q(A) unchanged. This establishes assertion (3) of Theorem 5.1.

(iii) If $A_i \cap \mathcal{C} \neq \emptyset$, then changing A_i to $A_i \cap \mathcal{C}$ will leave all the quantities (5.4) unchanged. Indeed, when $B \subset \mathcal{C}$, the condition $B \not\subset A_i$ is equivalent to the condition that $B \not\subset (A_i \cap \mathcal{C})$. And when $B \not\subset \mathcal{C}$, the quantity (5.4) is already zero and will remain so when the A_i are contracted. This establishes (4) of Theorem 5.1. ■

In light of the preceding proof, one might conjecture that the commonality function Q will remain unchanged as one varies the w_i and the A_i just so long as one leaves unchanged the sum

$$\sum_{\substack{i \\ A_i \cap \mathcal{C} = A}} w_i$$

for each non-empty subset A of $\mathcal{C}$. For this sum seems to be the effective weight of evidence focused on A. The correctness of this conjecture is confirmed in the next section.

§3. The Assessment of Evidence

Suppose the weights $w_1, \cdots, w_n$ and their foci $A_1, \cdots, A_n$ satisfy

$$\mathcal{C} \equiv \bigcap_{\substack{i \\ w_i = \infty}} A_i \neq \emptyset . \tag{5.5}$$

Then the function $w : 2^\Theta \to [0, \infty]$ given by

$$w(A) = \begin{cases} \sum_{\substack{i \\ A_i \cap \mathcal{C} = A}} w_i & \text{if } A \subset \mathcal{C},\ A \neq \emptyset, \text{ and } A \neq \mathcal{C} \\ \infty & \text{if } A = \mathcal{C} \\ 0 & \text{if } A = \emptyset \text{ or } A \not\subset \mathcal{C} \end{cases}$$

is called the *assessment of evidence* associated with the w_i.

The assessment of evidence is a useful substitute for the original weights because it captures precisely those aspects of the weights that can be recovered from the separable support function.

THEOREM 5.4. *Suppose the weights* $w_1, \cdots, w_n$ *are focused on* $A_1, \cdots, A_n \subset \Theta$ *and satisfy (5.5). Let* Q *be the commonality function determined by these weights, and let* w *be their assessment of evidence. Then*

(1) Q *may be obtained from* w:

$$Q(A) = K \exp\left(-\sum_{\substack{B \subset \Theta \\ A \not\subset B}} w(B)\right) \tag{5.6}$$

for all non-empty $A \subset \Theta$, *where*

$$K = \left(\sum_{\substack{B \subset \Theta \\ B \neq \emptyset}} (-1)^{|B|+1} \exp\left(-\sum_{\substack{C \subset \Theta \\ B \not\subset C}} w(C)\right)\right)^{-1}; \tag{5.7}$$

(2) w *may be obtained from* Q:

$$w(A) = \begin{cases} \sum_{\substack{B \subset \mathcal{C} \\ A \subset B}} (-1)^{|B-A|} \log Q(B) & \textit{if } A \subset \mathcal{C},\ A \neq \emptyset, \textit{ and } A \neq \mathcal{C} \\ \infty & \textit{if } A = \mathcal{C} \\ 0 & \textit{if } A = \emptyset \text{ or } A \not\subset \mathcal{C}, \end{cases}$$

where $\mathcal{C}$ *is the core of* Q.

To put the matter in different words, the assessment of evidence corresponds to the canonical decomposition of the separable support function.

THEOREM 5.5. *Suppose* $S: 2^{\Theta} \to [0,1]$ *is a non-vacuous separable support function, and suppose* w *is its assessment of evidence. Form a collection* $\mathcal{S}$ *of simple support functions by including, for each proper* $A \subset \Theta$ *such that* $w(A) > 0$, *the simple support function focused on* A *and based on the weight* $w(A)$. *Then* $\mathcal{S}$ *is the canonical decomposition of* S.

A function $w: 2^{\Theta} \to [0, \infty]$ will qualify as an assessment of evidence (i.e., it will be the assessment of evidence associated with some collection of weights) if and only if $w(\emptyset) = 0$, w assigns the value ∞ to exactly one subset $\mathcal{C}$ of Θ, and w assigns the value zero to every subset of Θ not contained in $\mathcal{C}$. Theorem 5.4 establishes that the assessments of evidence are in a one-to-one correspondence with the separable support functions.

§4. The Weight of Internal Conflict

To the extent that a separable support function lends support to pairs of disjoint subsets, or even to larger collections of subsets with empty intersections, it indicates internal conflict in the evidence. How can we best measure the amount of internal conflict that is indicated?

The amount of conflict actually present in the evidence underlying a separable support function S is best measured by $\mathrm{Con}(S_1, \cdots, S_n)$, where $S_1, \cdots, S_n$ are the simple support functions that actually combine to produce S. Since the decomposition of S into simple support functions will not be unique, knowledge of S alone will only afford us a minimum value for this true level of conflict. But this minimum value may aptly be termed the weight of internal conflict for S itself.

DEFINITION. The *weight of internal conflict* for a separable support function S is

(1) zero if S is a simple support function, and

(2) the infimum of the quantities $\mathrm{Con}(S_1,\cdots,S_n)$ for the various possible decompositions $S = S_1 \oplus \cdots \oplus S_n$ into simple support functions if S is not a simple support function.

I will denote the weight of internal conflict for a separable support function S by the symbol W together with an appropriate subscript. When attending to S itself I will write W_S; when attending to Q I will write W_Q; etc.

THEOREM 5.6. *Suppose* $S = 2^\Theta \to [0,1]$ *is a separable support function, and suppose* $\mathcal{C} \subset \Theta$ *is its core.*

(1) *If* w *denotes* S's *assessment of evidence, then*

$$W_S = -\log\left(\sum_{\substack{A\subset\Theta \\ A\neq\emptyset}} (-1)^{|A|+1} \exp\left(-\sum_{\substack{B\subset\Theta \\ A\not\subset B}} w(B)\right)\right). \tag{5.8}$$

(2) *If* S *is not simple, then* W_S *is equal to the weight of conflict among the simple support functions in its canonical decomposition.*

(3) *Suppose* $n \geq 2$, $S_1,\cdots,S_n$ *are simple support functions with foci intersecting* $\mathcal{C}$, *and* $S = S_1 \oplus \cdots \oplus S_n$. *Then* $W_S = \mathrm{Con}(S_1,\cdots,S_n)$.

(4) *Suppose* $2 \leq k < n$, $S_1,\cdots,S_k$ *are simple support functions with foci intersecting* $\mathcal{C}$, $S_{k+1},\cdots,S_n$ *are simple support functions with foci contained in* $\overline{\mathcal{C}}$, *and* $S = S_1 \oplus \cdots \oplus S_n$. *For each* i, *let* w_i *denote the weight of evidence underlying* S_i. *Then*

$$W_S = \mathrm{Con}(S_1,\cdots,S_n) - \sum_{i=k+1}^{n} w_i .$$

(5) *If* Q *denotes* S's *commonality function, then*

$$W_S = -\sum_{A\subset\mathcal{C}} (-1)^{|A|} \log Q(A) .$$

Comparing (5.7) and (5.8), we see that W_S is the logarithm of the renormalizing constant of Theorem 5.4.

Theorem 5.6 tells us that the conflict among a collection of simple support functions exceeds the internal conflict indicated by their orthogonal sum S only to the extent that some of the simple support functions directly impugn the core $\mathcal{C}$ of S. Such simple support functions are in direct conflict with the others, but are completely overwhelmed and obscured by them.

§5. The Impingement Function

We gain a measure of intuitive insight if we set

$$V(A) = \sum_{\substack{B \subset \Theta \\ A \not\subset B}} w(B) \tag{5.9}$$

for all $A \subset \Theta$ and rewrite (5.6) as

$$Q(A) = K e^{-V(A)}, \tag{5.10}$$

where

$$K = \left(\sum_{\substack{B \subset \Theta \\ B \neq \emptyset}} (-1)^{|B|+1} e^{-V(B)} \right)^{-1}. \tag{5.11}$$

Indeed, since each weight $w(B)$ impugns all those points of Θ not in its focus B and thus impugns part of A whenever $A \not\subset B$, $V(A)$ is the total weight impugning at least part of A. More felicitously: $V(A)$ is the total *weight of evidence impinging on* A. And (5.10) is intuitively sensible: less probability should be allowed to move freely in those subsets that have more evidence impinging on them.

I will call the function $V : 2^\Theta \to [0, \infty]$ that is defined by (5.9) the *impingement function.* (Notice that $V(\emptyset) = 0$.) This function obviously conveys exactly the same information as the assessment w. If we know w we can find V by (5.9); if we know V we can find w by (5.12) below, which may be obtained by combining (5.10) with the formula in Theorem 5.4 for obtaining w from Q.

THEOREM 5.7. *Suppose* V *is the impingement function for an assessment* $w : 2^\Theta \to [0, \infty]$. *Then the core* $\mathcal{C}$ *of* w *may be identified as the largest subset* A *of* Θ *for which* $V(A) < \infty$. *And*

$$w(A) = - \sum_{\substack{B \subset \mathcal{C} \\ A \subset B}} (-1)^{|B-A|} V(B) \tag{5.12}$$

for every proper non-empty subset A *of* Θ.

The constant given by (5.11) is the same as the constant given by (5.7). And by Theorem 5.6, the logarithm of this constant is equal to W_S. Since V conveys the same information as S, we may write W_V for W_S, and rewrite (5.10) as

$$-\log Q(A) = V(A) - W_V .$$

Intuitively: the weight of evidence impinging on A pushes down Q(A) only to the extent that it exceeds the weight of internal conflict in the evidence.

Using (5) of Theorem 5.6 to express W_S in terms of Q, we also obtain a formula for obtaining V from Q:

$$\begin{aligned} V(A) &= -\log Q(A) + W_Q \\ &= -\left(\log Q(A) + \sum_{A \subset \mathcal{C}} (-1)^{|B|} \log Q(B)\right), \end{aligned} \tag{5.13}$$

where $\mathcal{C}$ is the core of Q.

§6. The Weight-of-Conflict Conjecture

THE WEIGHT-OF-CONFLICT CONJECTURE. *If* Q_1 *and* Q_2 *are commonality functions for separable support functions over a frame* Θ *and obey*

$$Q_1(A) \le Q_2(A)$$

for all $A \subset \Theta$, *then* $W_{Q_1} \ge W_{Q_2}$.

This conjecture is difficult to motivate directly, but it has an immediate consequence with stronger intuitive appeal. Indeed, if V_1 and V_2 are two impingement functions over a frame Θ and Q_1 and Q_2 are their commonality functions, then (5.13) yields

$$V_1(A) = -\log Q_1(A) + W_{Q_1}$$

and

$$V_2(A) = -\log Q_2(A) + W_{Q_2}$$

for all $A \subset \Theta$. When $Q_1(A) > 0$, subtraction of the second from the first of these equations yields

$$V_1(A) - V_2(A) = \log \frac{Q_2(A)}{Q_1(A)} + (W_{Q_1} - W_{Q_2}) .$$

And we obtain the following theorem.

THEOREM 5.8. *Suppose* V_1 *and* V_2 *are two impingement functions over a frame* Θ, Q_1 *and* Q_2 *are their commonality functions, and*

$$Q_1(A) \leq Q_2(A)$$

for all $A \subset \Theta$. *Then if the weight-of-conflict conjecture is true, then*

$$V_1(A) \geq V_2(A)$$

for all $A \subset \Theta$.

The assertion in this theorem accords with the intuitive association of smaller commonality numbers with greater degrees of impingement, and it extends that association from the case where one compares different subsets of Θ under a fixed assessment to the case where one compares different assessments.

As we will see in Chapter 8, the weight-of-conflict conjecture plays an important role in the comparison of support functions over frames of varying refinement.

§7. Some Numerical Examples

The examples in this section contain no surprises, but they illustrate some of the possibilities not explicitly recognized in the preceding theoretical discussion.

Our first example shows that the converse of the assertion in Theorem 5.8 is not true. That is to say, two impingement functions V_1 and V_2 over Θ can satisfy

$$V_1(A) \geq V_2(A)$$

for all $A \subset \Theta$ without forcing their commonality functions to satisfy

$$Q_1(A) \leq Q_2(A)$$

for all $A \subset \Theta$.

EXAMPLE 5.1. *Counterexample*. Set $\Theta = \{a, b\}$, and define the two assessments w_1 and w_2 by setting

$$w_1(\{a\}) = \log 2 , \quad w_1(\{b\}) = \log 2$$

and

$$w_2(\{a\}) = \log 3 , \quad w_2(\{b\}) = \log 6 .$$

Table 5.1 shows the value for $\{a\}$ and $\{b\}$ of the corresponding impingement functions V_1, V_2 and the corresponding commonality functions Q_1, Q_2. Notice that $Q_2(\{b\}) > Q_1(\{b\})$, even though $V_1(A) \geq V_2(A)$ for all $A \subset \Theta$. The greater evidence against $\{b\}$ in the second assessment is more than counteracted by the greater evidence against $\{a\}$.

Table 5.1

	$V_i(\{a\})$	$V_i(\{b\})$	$V_i(\Theta)$	$Q_i(\{a\})$	$Q_i(\{b\})$	$Q_i(\Theta)$
i=1	log 2	log 2	log 4	$\frac{2}{3}$	$\frac{2}{3}$	$\frac{1}{3}$
i=2	log 6	log 3	log 18	$\frac{3}{8}$	$\frac{3}{4}$	$\frac{1}{8}$

■

The two following examples should remind us of the complexity of the relation between commonality numbers and degrees of support, for they show us that smaller commonality numbers may be compatible with either smaller or greater degrees of support. Both examples exhibit pairs Q_1, Q_2 of commonality functions over a frame Θ such that $Q_1(A) \leq Q_2(A)$ for all $A \subset \Theta$. The corresponding support functions S_1 and S_2 in Example 5.2 obey $S_1(A) \leq S_2(A)$ for all $A \subset \Theta$ and $S_1(A) < S_2(A)$ for some $A \subset \Theta$; those in Example 5.3 obey $S_2(A) \leq S_1(A)$ for all $A \subset \Theta$ and $S_2(A) < S_1(A)$ for some $A \subset \Theta$.

EXAMPLE 5.2. *More Precise Evidence Can Be More Efficient.* Set $\Theta = \{a, b, c\}$, and define assessments w_1 and w_2 as follows:

$$w_1(\{a\}) = \log 2 ,$$

$w_1(\Theta) = \infty$, and $w_1(A) = 0$ for all other $A \subset \Theta$. And

$$w_2(\{a, b\}) = w_2(\{a, c\}) = \log 2 ,$$

$w_2(\Theta) = \infty$, and $w_2(A) = 0$ for all other $A \subset \Theta$. The evidence represented by the assessment w_1 has half the total weight of that represented by w_2, but it is more precise.

Table 5.2

A	$e^{V_1(A)}$	$e^{V_2(A)}$	$Q_1(A)$	$Q_2(A)$	$S_1(A)$	$S_2(A)$
$\{a\}$	0	0	1	1	1/2	1/4
$\{b\}$	2	2	1/2	1/2	0	0
$\{c\}$	2	2	1/2	1/2	0	0
$\{a,b\}$	2	2	1/2	1/2	1/2	1/2
$\{a,c\}$	2	2	1/2	1/2	1/2	1/2
$\{c,d\}$	2	4	1/2	1/4	0	0
Θ	2	4	1/2	1/4	1	1

Table 5.2 exhibits the corresponding impingement functions V_1, V_2, commonality functions Q_1, Q_2, and separable support functions S_1, S_2. Notice that the more precise assessment w_1 yields the greatest degree of support for $\{a\}$. ■

EXAMPLE 5.3. *Evidence on Two Sides of a Dichotomy.* Set $\Theta = \{a, b, c, d\}$, and define assessments w_1 and w_2 as follows:

$$w_1(\{a, b\}) = w_1(\{c, d\}) = \log 2 ,$$

$w_1(\Theta) = \infty$ and $w_1(A) = 0$ for all other $A \subset \Theta$. And

$$w_2(\{a, c\}) = w_2(\{b, d\}) = \log 3 ,$$

$$w_2(\{a, d\}) = \log 6 ,$$

$w_2(\Theta) = \infty$ and $w_2(A) = 0$ for all other $A \subset \Theta$.

Table 5.3

A	$e^{V_1(A)}$	$e^{V_2(A)}$	$Q_1(A)$	$Q_2(A)$	$S_1(A)$	$S_2(A)$
{a}	2	3	2/3	3/5	0	1/3
{b}	2	18	2/3	1/10	0	0
{c}	2	18	2/3	1/10	0	0
{d}	2	3	2/3	3/5	0	1/3
{a,b}	2	54	2/3	1/30	1/3	1/3
{a,c}	4	18	1/3	1/10	0	2/5
{a,d}	4	9	1/3	1/5	0	5/6
{b,c}	4	54	1/3	1/30	0	0
{b,d}	4	18	1/3	1/10	0	2/5
{c,d}	2	54	2/3	1/30	1/3	1/3
{a,b,c}	4	54	1/3	1/30	1/3	2/5
{a,b,d}	4	54	1/3	1/30	1/3	9/10
{a,c,d}	4	54	1/3	1/30	1/3	9/10
{b,c,d}	4	54	1/3	1/30	1/3	2/5
Θ	4	54	1/3	1/30	1	1

Table 5.3 shows the values that w_1 and w_2 yield for the corresponding functions V_1, V_2, Q_1, Q_2, S_1 and S_2. Notice that both w_1 and w_2 have the same impact on the dichotomy $\{a,b\}$ versus $\{c,d\}$; both sides of this dichotomy receive a degree of support of $\frac{1}{3}$ under either assessment. ■

§8. Mathematical Appendix

The reader will notice that the first theorem proven in this appendix is Theorem 5.3. Theorem 5.1 is proven in §2 above, and Theorem 5.2 is listed as a corollary to Lemma 5.4 below.

LEMMA 5.1. *Suppose* $\mathcal{C}$ *is a finite set and* f *and* g *are functions on* $2^{\mathcal{C}}$. *Then*

$$f(A) = \sum_{\substack{B \subset \mathcal{C} \\ A \not\subset B}} g(B) \tag{5.14}$$

for all non-empty $A \subset \mathcal{C}$ *if and only if*

$$g(A) = -\sum_{\substack{B \subset \mathcal{C} \\ A \subset B}} (-1)^{|B-A|} f(B) \tag{5.15}$$

for all proper non-empty subsets A *of* $\mathcal{C}$.

Proof of Lemma 5.1.

(i) Suppose (5.14) holds for all non-empty $A \subset \mathcal{C}$. Then using Lemma 2.2, we find that

$$-\sum_{\substack{B \subset \mathcal{C} \\ A \subset B}} (-1)^{|B-A|} f(B) = -\sum_{\substack{B \subset \mathcal{C} \\ A \subset B}} (-1)^{|B-A|} \sum_{\substack{C \subset \mathcal{C} \\ B \not\subset C}} g(C)$$

$$= (-1)^{|A|+1} \sum_{C \subset \mathcal{C}} g(C) \sum_{\substack{B \subset \mathcal{C} \\ A \subset B \not\subset C}} (-1)^{|B|}$$

$$= (-1)^{|A|+1} \sum_{C \subset \mathcal{C}} g(C) \left(\sum_{\substack{B \subset \mathcal{C} \\ A \subset B}} (-1)^{|B|} - \sum_{\substack{B \subset \mathcal{C} \\ A \subset B \subset C}} (-1)^{|B|} \right)$$

$$= (-1)^{|A|+1} g(A) ,$$

provided A is a proper non-empty subset of $\mathcal{C}$.

(ii) Suppose (5.15) holds for all proper non-empty subsets A of $\mathcal{C}$. Then using Lemmas 2.1 and 2.2, we find that

$$\sum_{\substack{B \subset \mathcal{C} \\ A \not\subset B}} g(B) = - \sum_{\substack{B \subset \mathcal{C} \\ A \not\subset B}} \sum_{\substack{C \subset \mathcal{C} \\ B \subset C}} (-1)^{|C-B|} f(C)$$

$$= - \sum_{\substack{C \subset \mathcal{C} \\ C \neq \emptyset}} (-1)^{|C|} f(C) \sum_{\substack{B \subset \mathcal{C} \\ A \not\subset B \subset C}} (-1)^{|B|}$$

$$= - \sum_{\substack{C \subset \mathcal{C} \\ C \neq \emptyset}} (-1)^{|C|} f(C) \left(\sum_{\substack{B \subset \mathcal{C} \\ B \subset C}} (-1)^{|B|} - \sum_{\substack{B \subset \mathcal{C} \\ A \subset B \subset C}} (-1)^{|B|} \right).$$

$$= - (-1)^{|A|} f(A) (-(-1)^{|A|})$$

$$= f(A) ,$$

provided A is a non-empty subset of $\mathcal{C}$. ∎

LEMMA 5.2. *Suppose* $\mathcal{C}$ *is a finite set and* $Q : (2^{\mathcal{C}} - \{\emptyset\}) \to (0, \infty)$ *and* $w : (2^{\mathcal{C}} - \{\emptyset, \mathcal{C}\}) \to [0, \infty)$ *are two functions. Then*

$$Q(A) = K \exp\left(- \sum_{\substack{B \subset \mathcal{C} \\ A \not\subset B \neq \emptyset}} w(B) \right) \qquad (5.16)$$

for all $A \in (2^{\mathcal{C}} - \{\emptyset\})$ *and some positive constant* K *if and only if*

$$w(A) = \sum_{\substack{B \subset \mathcal{C} \\ A \subset B}} (-1)^{|B-A|} \log Q(B) \tag{5.17}$$

for all $A \in (2^{\mathcal{C}} - \{\emptyset, \mathcal{C}\})$.

Proof of Lemma 5.2.

(i) Suppose (5.16) holds for all $A \in (2^{\mathcal{C}} - \{\emptyset\})$. Let f be a function on $2^{\mathcal{C}}$ that satisfies $f(A) = -\log Q(A)$ for all $A \in (2^{\mathcal{C}} - \{\emptyset\})$, and let g be a function on $2^{\mathcal{C}}$ that satisfies $g(A) = w(A)$ for all $A \in (2^{\mathcal{C}} - \{\emptyset, \mathcal{C}\})$ and $g(\emptyset) = -\log K$. Then

$$f(A) = -\log Q(A) = -\log K + \sum_{\substack{B \subset \mathcal{C} \\ A \not\subset B \neq \emptyset}} w(B) = \sum_{\substack{B \subset \mathcal{C} \\ A \not\subset B}} g(B)$$

for all non-empty $A \subset \mathcal{C}$. It follows by Lemma 5.1 that for all $A \in (2^{\mathcal{C}} - \{\emptyset, \mathcal{C}\})$

$$w(A) = g(A) = -\sum_{\substack{B \subset \mathcal{C} \\ A \subset B}} (-1)^{|B-A|} f(B) = \sum_{\substack{B \subset \mathcal{C} \\ A \subset B}} (-1)^{|B-A|} \log Q(B) .$$

(ii) Suppose (5.17) holds for all $A \in (2^{\mathcal{C}} - \{\emptyset, \mathcal{C}\})$. Let f be a function on $2^{\mathcal{C}}$ satisfying $f(A) = -\log Q(A)$ for all $A \in (2^{\mathcal{C}} - \{\emptyset\})$, and let g be a function on $2^{\mathcal{C}}$ satisfying $g(A) = w(A)$ for all $A \in (2^{\mathcal{C}} - \{\emptyset, \mathcal{C}\})$. Then (5.17) immediately becomes (5.15), and by Lemma 5.1,

$$w(A) = -\log Q(A) = \sum_{\substack{B \subset \mathcal{C} \\ A \not\subset B}} g(B) = g(\emptyset) + \sum_{\substack{B \subset \mathcal{C} \\ A \not\subset B \neq \emptyset}} w(B)$$

for all $A \in (2^{\mathcal{C}} - \{\emptyset\})$. Hence (5.16) holds for all $A \in (2^{\mathcal{C}} - \{\emptyset\})$, where

$$K = e^{-g(\emptyset)} > 0 . \blacksquare$$

Proof of Theorem 5.3. Since S_i assigns the basic probability number $1 - e^{-w_i}$ to A_i and the basic probability number e^{-w_i} to Θ, its commonality function Q_i is given by

$$Q_i(A) = \begin{cases} 1 & \text{if } A \subset A_i \\ e^{-w_i} & \text{if } A \not\subset A_i , \end{cases}$$

and its core $\mathcal{C}_i$ is given by

$$\mathcal{C}_i = \begin{cases} A_i & \text{if } w_i = \infty \\ \Theta & \text{if } w_i < \infty . \end{cases}$$

So the orthogonal sum $S_1 \oplus \cdots \oplus S_n$ exists provided only that

$$\mathcal{C}_1 \cap \cdots \cap \mathcal{C}_n = \bigcap_{\substack{i \\ w_i = \infty}} A_i = \mathcal{C} \neq \emptyset ;$$

and if it does exist, its core is $\mathcal{C}$, and its commonality function Q is given by

$$Q(A) = K \prod_{i=1}^{n} Q_i(A)$$

$$= K \exp\left(- \sum_{\substack{i \\ A \not\subset A_i}} w_i \right)$$

for all non-empty $A \subset \Theta$. The expression for the constant K then follows from (2.6).

By (5.3), $\mathcal{C}$ is contained in all the A_i for which $w_i = \infty$ and contains any other subset A of Θ that is contained in all those A_i. Hence there exists i such that $w_i = \infty$ and $A \not\subset A_i$ if and only if $A \not\subset \mathcal{C}$. It follows by (5.1) that $Q(A) > 0$ if and only if $A \subset \mathcal{C}$. ■

LEMMA 5.3. *Suppose that for* $i = 1, \cdots, n$, S_i *is a simple support function focused on* A_i *and based on the weight of evidence* w_i. *And suppose that* $S = S_1 \oplus \cdots \oplus S_n$ *exists and has core* $\mathcal{C}$ *and assessment* w. *Then*

$$\sum_{\substack{B \subset \Theta \\ A \not\subset B}} w(B) = \sum_{\substack{i \\ A \not\subset A_i}} w_i - \sum_{\substack{i \\ A_i \subset \overline{\mathcal{C}}}} w_i$$

for all $A \subset \Theta$.

Proof of Lemma 5.3. If A is a non-empty subset of $\mathcal{C}$, then

$$\sum_{\substack{B \subset \Theta \\ A \not\subset B}} w(B) = \sum_{\substack{B \subset \mathcal{C} \\ A \not\subset B \neq \emptyset}} w(B) = \sum_{\substack{B \subset \mathcal{C} \\ A \not\subset B \neq \emptyset}} \sum_{\substack{i \\ A_i \cap \mathcal{C} = B}} w_i$$

$$= \sum_{\substack{i \\ A \not\subset A_i \not\subset \overline{\mathcal{C}}}} w_i = \sum_{\substack{i \\ A \not\subset A_i}} w_i - \sum_{\substack{i \\ A_i \subset \overline{\mathcal{C}}}} w_i ,$$

the last step being possible only because $w_i < \infty$ for all i such that $A_i \subset \overline{\mathcal{C}}$. The same result can be obtained when $A \not\subset \mathcal{C}$, for then

$$\sum_{\substack{B \subset \Theta \\ A \not\subset B}} w(B) \geq w(\mathcal{C}) = \infty = \sum_{\substack{i \\ A \not\subset A_i}} w_i$$

$$= \sum_{\substack{i \\ A \not\subset A_i}} w_i - \sum_{\substack{i \\ A_i \subset \overline{\mathcal{C}}}} w_i . \blacksquare$$

Proof of Theorem 5.4.

(1) Using the preceding lemma, we find that this theorem's expression for $Q(A)$ becomes

$$\frac{\exp\left(-\sum\limits_{\substack{i \\ A \not\subset A_i}} w_i + \sum\limits_{\substack{i \\ A_i \subset \overline{\mathcal{C}}}} w_i\right)}{\sum\limits_{\substack{B \subset \Theta \\ B \neq \emptyset}} (-1)^{|B|+1} \exp\left(-\sum\limits_{\substack{i \\ B \not\subset A_i}} w_i + \sum\limits_{\substack{i \\ A_i \subset \overline{\mathcal{C}}}} w_i\right)} = \frac{\exp\left(-\sum\limits_{\substack{i \\ A \not\subset A_i}} w_i\right)}{\sum\limits_{\substack{B \subset \Theta \\ B \neq \emptyset}} (-1)^{|B|+1} \exp\left(-\sum\limits_{\substack{i \\ B \not\subset A_i}} w_i\right)},$$

which is indeed equal to $Q(A)$ by (5.1).

(2) Since any assessment w satisfies $w(\emptyset) = 0$, (5.6) is equivalent to (5.16) and hence implies (5.17) by Lemma 5.2. This confirms the theorem's expression for $w(A)$ for the case where A is a proper non-empty subset of $\mathcal{C}$; the expressions for the other cases hold by the definition of an assessment. ■

LEMMA 5.4. *Suppose* S *is a separable support function over* Θ *and* w *is its assessment of evidence. Then* S *is vacuous if and only if* $w(A)=0$ *for all proper subsets* A *of* Θ.

Suppose S *is non-vacuous, and let* $A_1,\cdots,A_n$ *enumerate all the (distinct) proper subsets of* Θ *to which* w *assigns a positive value. For each* A_i, $i = 1,\cdots,n$, *let* S_i *be the simple support function focused on* A_i *with* $S_i(A_i) = 1 - e^{-w(A_i)}$. *Then the collection* $S_1,\cdots,S_n$ *satisfies the four conditions of Theorem 5.2, and it is the only collection satisfying them.*

Proof of Lemma 5.4.

(i) A belief function S over Θ is vacuous if and only if its commonality function Q satisfies $Q(A) = 1$ for all $A\subset\Theta$. So a glance at (5.6) shows that S is vacuous when $w(A) = 0$ for all proper subsets of Θ. On the other hand, if S is vacuous, then the requirement that $Q(A) = 1$ for all $A\subset\Theta$ translates by (5.6) into the requirement that the sum

$$\sum_{\substack{B\subset\Theta \\ A\not\subset B}} w(B)$$

be the same for all non-empty $A\subset\Theta$. In particular,

$$\sum_{\substack{B\subset\Theta \\ A\not\subset B}} w(B) = \sum_{\substack{B\subset\Theta \\ \Theta\not\subset B}} w(B) = \sum_{\substack{B\subset\Theta \\ B\neq\Theta}} w(B)$$

for all non-empty $A\subset\Theta$. So when A is a proper non-empty subset of Θ, one must have

$$0 = \sum_{\substack{B \subset \Theta \\ B \neq \Theta}} w(B) - \sum_{\substack{B \subset \Theta \\ A \not\subset B}} w(B) = \sum_{\substack{B \subset \Theta \\ A \subset B \neq \Theta}} w(B) \geq w(A) \,,$$

or $w(A) = 0$. And of course $w(\emptyset) = 0$ in any case.

(ii) Now let us verify that the collection $S_1, \cdots, S_n$ satisfies the conditions of Theorem 5.2. Since S is non-vacuous, the first paragraph of the present lemma assures us that condition (1) is satisfied. Conditions (3) and (4) are obvious from the way $S_1, \cdots, S_n$ are constructed; so we need only attend to condition (2). By (5.1) and (5.6), it suffices to show that

$$\sum_{\substack{i \\ A \not\subset A_i}} w(A_i) = \sum_{\substack{B \subset \Theta \\ A \not\subset B}} w(B)$$

for all non-empty $A \subset \Theta$. But this is true because the A_i include all the proper $B \subset \Theta$ for which $w(B) > 0$.

(iii) Now we must show that no other collection of non-vacuous simple support functions can satisfy the four conditions of Theorem 5.2. To this end, suppose that for $i = 1, \cdots, m$, S'_i is a simple support function focused on $A'_i \subset \Theta$ with $S(A'_i) = 1 - e^{-w'_i}$, and suppose that the collection $S'_1, \cdots, S'_m$ satisfies four conditions:

(1) $m \geq 1$.

(2) $S = S'_1$ if $m = 1$ and $S = S'_1 \oplus \cdots \oplus S'_m$ if $m > 1$.

(3) $A'_i \subset \mathcal{C}$ for $i = 1, \cdots, m$.

(4) If $1 \leq i < j \leq m$, then $A'_i \neq A'_j$.

Since the assessment of evidence associated with a separable support function is unique, (2) means that w is the assessment associated with the w'_i. Hence

$$w(A) = \sum_{\substack{i \\ A'_i \cap \mathcal{C} = A}} w'_i$$

for all non-empty proper subsets A of $\mathcal{C}$. By (3), this becomes

$$w(A) = \sum_{\substack{i \\ A'_i = A}} w'_i$$

for all non-empty proper subsets A of $\mathcal{C}$. And by (4), it follows that the collection of those S'_i whose foci are proper subsets of $\mathcal{C}$ is identical with the collection of those S_i whose foci are proper subsets of $\mathcal{C}$. If $\mathcal{C} = \Theta$, this means that the collection $S'_1, \cdots, S'_m$ is identical with the collection $S_1, \cdots, S_n$.

If $\mathcal{C}$ is a proper subset of Θ, then the S_i include exactly one simple support function whose focus is not a proper subset of $\mathcal{C}$, namely one focusing an infinite weight on $\mathcal{C}$; and we must show that this is true of the S'_i too. But by (3), $\mathcal{C}$ is the only subset of Θ besides a proper subset of $\mathcal{C}$ that one of the S'_i could be focused on. And since

$$\mathcal{C} = \bigcap_{\substack{i \\ w'_i = \infty}} A'_i ,$$

there must indeed be an i for which $A'_i = \mathcal{C}$ and $w'_i = \infty$. ∎

Theorems 5.2 and 5.5 are immediate corollaries of this lemma.

Proof of Theorem 5.6.

Let us first consider the case where S is simple (or vacuous). Say S is a simple support function focused on $A_1 \subset \Theta$ and based on the weight of evidence w_1. Then S's assessment w is given by $w(A_1) = w_1$ and $w(A) = 0$ for all other proper subsets of Θ. And

$$\sum_{\substack{B \subset \Theta \\ A \not\subset B}} w(B) = \begin{cases} 0 & \text{if } A \subset A_1 \\ w_1 & \text{if } A \not\subset A_1 , \end{cases}$$

and

$$-\log\left(\sum_{\substack{A\subset\Theta\\A\neq\emptyset}}(-1)^{|A|+1}\exp\left(-\sum_{\substack{B\subset\Theta\\A\not\subset B}}w(B)\right)\right) = -\log\left(\sum_{\substack{A\subset\Theta\\\emptyset\neq A\subset A_1}}(-1)^{|A|+1}+e^{-w_1}\sum_{\substack{A\subset\Theta\\A\not\subset A_1}}(-1)^{|A|+1}\right)$$

$$= -\log 1$$

$$= 0 .$$

Hence (5.8) holds when S is simple.

Suppose now that for $i = 1,\cdots,n$, S_i is a simple support function focused on $A_i \subset \Theta$ and based on the weight w_i. And suppose $S = S_1 \oplus \cdots \oplus S_n$. Then by Lemma 5.3,

$$\sum_{\substack{i\\A\not\subset A_i}} w_i = \sum_{\substack{B\subset\Theta\\A\not\subset B}} w(B) + \sum_{\substack{i\\A_i\subset\bar{\mathcal{C}}}} w_i ,$$

and hence

$$\mathrm{Con}(S_1,\cdots,S_n) = -\log\left(\sum_{\substack{A\subset\Theta\\A\neq\emptyset}}(-1)^{|A|+1}\exp\left(-\sum_{\substack{i\\A\not\subset A_i}}w_i\right)\right)$$

$$= -\log\left(\sum_{\substack{A\subset\Theta\\A\neq\emptyset}}(-1)^{|A|+1}\exp\left(-\sum_{\substack{B\subset\Theta\\A\not\subset B}}w(B)\right)\right) + \sum_{\substack{i\\A_i\subset\bar{\mathcal{C}}}} w_i .$$

Now the last term is zero if the decomposition $S = S_1 \oplus \cdots \oplus S_n$ is canonical, or more generally, if all the A_i intersect $\mathcal{C}$. This fact immediately yields assertions (1), (2), (3) and (4) of our theorem.

Using the expression for $Q(A)$ in (5.6),

$$-\sum_{A\subset\mathcal{C}}(-1)^{|A|}\log Q(A) = \log K + \sum_{\substack{A\subset\mathcal{C}\\A\neq\emptyset}}(-1)^{|A|}\sum_{\substack{B\subset\Theta\\A\not\subset B}}w(B)$$

$$= \log K + \sum_{B\subset\Theta}w(B)\sum_{\substack{A\subset\mathcal{C}\\A\not\subset B}}(-1)^{|A|} = \log K ,$$

where K is the constant given by (5.7). But comparison of (5.7) and (5.8) reveals that $\log K = W_S$. ■

Proof of Theorem 5.7. Since $w(\mathcal{C}) = \infty$ and $w(B) = 0$ for all other $B \subset \Theta$,

$$\sum_{\substack{B \subset \Theta \\ B \not\subset A}} w(B) = \infty$$

if and only if $A \not\subset \mathcal{C}$. So $V(A) < \infty$ if and only if $A \subset \mathcal{C}$ – i.e., $\mathcal{C}$ is the largest subset A of Θ for which $V(A) < \infty$.

By Theorem 5.4,

$$w(A) = \sum_{\substack{B \subset \mathcal{C} \\ A \subset B}} (-1)^{|B-A|} \log Q(B)$$

for all proper non-empty subsets A of $\mathcal{C}$. Substituting $K\, e^{-V(B)}$ for $\log Q(B)$ in this equation yields

$$w(A) = -\sum_{\substack{B \subset \mathcal{C} \\ A \subset B}} (-1)^{|B-A|} V(B) + (-1)^{|A|} \sum_{\substack{B \subset \mathcal{C} \\ A \subset B}} (-1)^{|B|} \log K \,.$$

And this becomes (5.12) when we notice that the second term is zero by Lemma 2.2. ■

Proof of Theorem 5.8. If we assume that $Q_1(A) \leq Q_2(A)$ for all $A \subset \Theta$, then the weight-of-conflict conjecture tells us that $W_{Q_1} - W_{Q_2} \geq 0$. Hence the formula

$$V_1(A) - V_2(A) = \log \frac{Q_2(A)}{Q_1(A)} + (W_{Q_1} - W_{Q_2})$$

derived in the text preceding the theorem, tells us that

$$V_1(A) \geq V_2(A) \tag{5.18}$$

for all non-empty A for which $Q_1(A) > 0$. But since $V_1(\emptyset) = V_2(\emptyset) = 0$, (5.18) necessarily holds when A is empty. And if $Q_1(A) = 0$, then $V_1(A) = \infty$ by (5.13), and again (5.18) must hold. ■

CHAPTER 6. COMPATIBLE FRAMES OF DISCERNMENT

> ··· ogni idea che non tenga conto della relatività e arbitrarietà e provvisorietà di tale arresto nella suddivisione, che pensi ad essi come "indivisibili" o come "meno suddivisibili" o communque diversi da tutti gli altri eventi, è infondata e ingannevole.
>
> BRUNO DE FINETTI (b. 1906)

Being an exact and formal list of possibilities, a single frame of discernment can embody only a small subset of the immense collection of concepts and distinctions that any thinker can call to his aid. And by using different concepts or emphasizing different distinctions, a thinker can vary the frame with which he approaches a particular instance of probable reasoning. Hence a mathematical theory of evidence would do well to go beyond the study of single frames and explore the relations among different frames.

In this chapter, we study frames that are different but compatible – frames which emphasize different distinctions and thus differ in the direction and degree of resolution of their attention, but which do not employ contradictory or incompatible concepts. In Chapter 12, we will consider the more difficult problem of comparing frames that are incompatible.

One frame is certainly compatible with another if it can be obtained from it by introducing new distinctions – i.e., by analyzing or splitting some of its possibilities into finer possibilities. In such a case, the frame of the finer analysis is called a *refinement*; the other is called a *coarsening*. In general, two frames can be compatible without either being a refinement of the other, but the requirement that the distinctions and concepts underlying the two be compatible means that the two must have a common refinement. Hence a *family of compatible frames* will be a collection of frames any pair of which has a common refinement. In §1 and §2 below I develop a mathematical description of the relation between coarsenings and their refinements; in §3 and §4 I investigate the mathematical structure of a family of compatible frames.

In §5 I turn to the question of how belief functions over compatible frames will be related when they express the same opinion, and in §6 I formulate the standard but important notion of *independence* for compatible frames.

§1. Refinements and Coarsenings

The idea that one frame of discernment Ω is obtained from another frame of discernment Θ by splitting some or all of the elements of Θ may be represented mathematically by specifying, for each $\theta \in \Theta$, the subset $\omega(\{\theta\})$ of Ω consisting of those possibilities into which θ has been split. For this representation to be sensible, we need only require that the sets $\omega(\{\theta\})$ should constitute a disjoint partition of Ω. In other words,

(1) $\omega(\{\theta\}) \neq \emptyset$ for all $\theta \in \Theta$,

(2) $\omega(\{\theta\}) \cap \omega(\{\theta'\}) = \emptyset$ if $\theta \neq \theta'$,

(3) $\bigcup_{\theta \in \Theta} \omega(\{\theta\}) = \Omega$.

Given such a disjoint partition $\omega(\{\theta\})$, we may set

$$\omega(A) = \bigcup_{\theta \in A} \omega(\{\theta\}) \tag{6.1}$$

for each $A \subset \Theta$; $\omega(A)$ will consist of all the possibilities in Ω that are obtained by splitting the elements of A, and the mapping $\omega : 2^{\Theta} \to 2^{\Omega}$ that is thus defined will provide a thorough description of the splitting.

I call such a mapping ω a *refining*. In other words, $\omega : 2^{\Theta} \to 2^{\Omega}$ is a refining whenever Θ and Ω are finite sets, the sets $\omega(\{\theta\})$ constitute a disjoint partition of Ω, and the sets $\omega(A)$ are given in terms of the

$\omega(\{\theta\})$ by (6.1). Whenever $\omega : 2^{\Theta} \to 2^{\Omega}$ is a refining, I call Ω a *refinement* of Θ, and I call Θ a *coarsening* of Ω.

Notice that the propositions discerned by a frame of discernment Ω include all those propositions discerned by a coarsening Θ. Indeed, the proposition represented by a subset $A \subset \Theta$ will also be represented by the subset $\omega(A) \subset \Omega$. So when 2^{Θ} and 2^{Ω} are thought of as sets of propositions rather than as sets of subsets, 2^{Θ} must be considered a subset of 2^{Ω}.

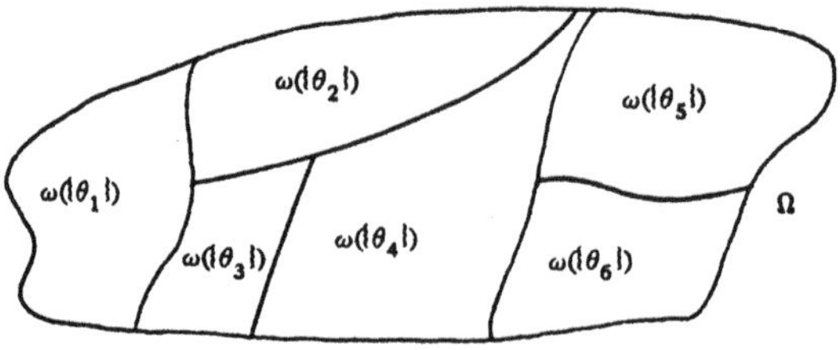

Figure 6.1. A coarsening $\Theta = \{\theta_1, \cdots, \theta_6\}$ of a frame Ω, depicted as a disjoint partition of Ω.

The following theorem exhibits some of the properties of refinings:

THEOREM 6.1. *Suppose* $\omega : 2^{\Theta} \to 2^{\Omega}$ *is a refining. Then*

(1) ω *is one-to-one*;

(2) $\omega(\emptyset) = \emptyset$;

(3) $\omega(\Theta) = \Omega$;

(4) $\omega(A \cup B) = \omega(A) \cup \omega(B)$ *for all* $A, B \subset \Theta$;

(5) $\omega(\overline{A}) = \overline{\omega(A)}$ *for all* $A \subset \Theta$;

(6) $\omega(A \cap B) = \omega(A) \cap \omega(B)$ *for all* $A, B \subset \Theta$;

(7) *if* $A, B \subset \Theta$, *then* $\omega(A) \subset \omega(B)$ *if and only if* $A \subset B$;

(8) *if* $A, B \subset \Theta$, *then* $\omega(A) \cap \omega(B) = \emptyset$ *if and only if* $A \cap B = \emptyset$.

EXAMPLE 6.1. *The Burglary of the Sweetshop.* In Example 4.1, we used the frame of discernment

$$\Theta = \{LI, LO, RI, RO\} ,$$

where LI denoted the possibility that the thief is a left-handed insider, etc. We may refine Θ by introducing the question of

Sherlock Holmes' age. Set

$$\Omega = \{li_1, li_2, lo, ri, ro\},$$

where li_1 denotes the possibility that the thief is a left-handed insider and that Holmes is under thirty, while li_2 denotes the possibility that the thief is a left-handed insider and that Holmes is thirty or older. Then the refining $\omega : 2^\Theta \to 2^\Omega$ is given by

$$\omega(\{LI\}) = \{li_1, li_2\},$$

$$\omega(\{LO\}) = \{lo\},$$

$$\omega(\{RI\}) = \{ri\},$$

and

$$\omega(\{RO\}) = \{ro\}$$

for singletons and by (6.1) for non-singletons. ■

§2. The Inner and Outer Reductions

A refining $\omega : 2^\Theta \to 2^\Omega$ is not, in general, onto; there are usually subsets A of Ω that are not discerned by Θ and hence are not equal to $\omega(B)$ for any $B \subset \Theta$. Nonetheless, there are two interesting and useful ways of associating a subset of Θ with each subset A of Ω.

With every refining $\omega : 2^\Theta \to 2^\Omega$, we may associate an *inner reduction* and an *outer reduction*. The inner reduction is the mapping $\underline{\theta} : 2^\Omega \to 2^\Theta$ given by

$$\underline{\theta}(A) = \{\theta \in \Theta \mid \omega(\{\theta\}) \subset A\}$$

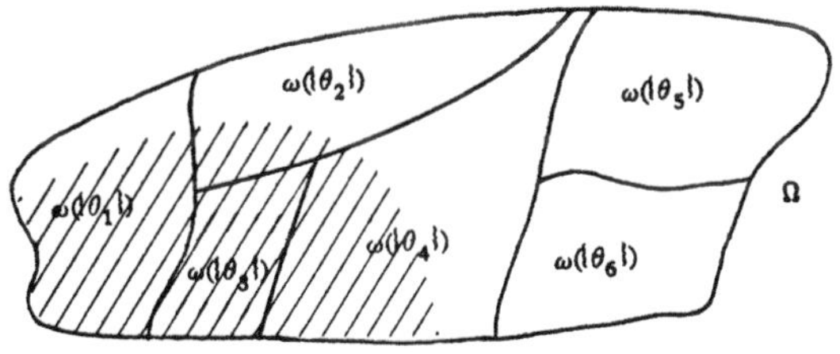

Figure 6.2.

for all $A \subset \Omega$, and the outer reduction is the mapping $\overline{\theta}: 2^{\Omega} \rightarrow 2^{\Theta}$ given by

$$\overline{\theta}(A) = \{\theta \in \Theta | \omega(\{\theta\}) \cap A \neq \emptyset\}$$

for all $A \subset \Omega$.

Figure 6.2 allows us to observe $\underline{\theta}(A)$ and $\overline{\theta}(A)$ for certain subsets A of the frame Ω of Figure 6.1. If we take A to be the shaded subset of Ω in the figure, then $\underline{\theta}(A) = \{\theta_3\}$ and $\overline{\theta}(A) = \{\theta_1, \theta_2, \theta_3, \theta_4\}$. If we take A to be the unshaded subset, then $\underline{\theta}(A) = \{\theta_5, \theta_6\}$ and $\overline{\theta}(A) = \{\theta_1, \theta_2, \theta_4, \theta_5, \theta_6\}$.

The following theorem will enable us to give $\underline{\theta}(A)$ and $\overline{\theta}(A)$ a more intuitive interpretation.

THEOREM 6.2. *Suppose* $\omega: 2^{\Theta} \rightarrow 2^{\Omega}$ *is a refining,* $A \subset \Omega$, *and* $B \subset \Theta$. *Let* $\underline{\theta}$ *and* $\overline{\theta}$ *be the inner and outer reductions for* ω. *Then* $\omega(B) \subset A$ *if and only if* $B \subset \underline{\theta}(A)$. *And* $A \subset \omega(B)$ *if and only if* $\overline{\theta}(A) \subset B$.

Now since B and $\omega(B)$ correspond to the same proposition, the relation $\omega(B) \subset A$ can be interpreted by saying that B implies A. So Theorem 6.2 tells us that B implies A if and only if $B \subset \underline{\theta}(A)$. In other words, $\underline{\theta}(A)$ implies A, and no other subset of Θ implies A unless it is a subset of $\underline{\theta}(A)$; $\underline{\theta}(A)$ *is the largest subset of* Θ *that implies* A. Similarly, A implies B if and only if $B \subset \overline{\theta}(A)$; A implies $\overline{\theta}(A)$ and implies no other subsets of Θ except those containing $\overline{\theta}(A)$; $\overline{\theta}(A)$ *is the smallest subset of* Θ *that is implied by* A.

The following theorem exhibits some of the properties of inner and outer reductions.

THEOREM 6.3. *Suppose* $\omega: 2^{\Theta} \rightarrow 2^{\Omega}$ *is a refining, and let* $\underline{\theta}$ *and* $\overline{\theta}$ *denote its inner and outer reductions. Then*

(1) $\underline{\theta}(A) \subset \overline{\theta}(A)$ *for all* $A \subset \Omega$;

(2) $\underline{\theta}(\omega(A)) = \overline{\theta}(\omega(A)) = A$ *for all* $A \subset \Theta$;

(3) $\omega(\underline{\theta}(A)) \subset A$ *for all* $A \subset \Omega$;

(4) $A \subset \omega(\bar{\theta}(A))$ *for all* $A \subset \Omega$;

(5) $\underline{\theta}(\emptyset) = \bar{\theta}(\emptyset) = \emptyset$;

(6) $\underline{\theta}(\Omega) = \bar{\theta}(\Omega) = \Theta$;

(7) *if* $A \subset B$, *then* $\underline{\theta}(A) \subset \underline{\theta}(B)$, *and* $\bar{\theta}(A) \subset \bar{\theta}(B)$;

(8) $\underline{\theta}(A \cap B) = \underline{\theta}(A) \cap \underline{\theta}(B)$ *for all* $A, B \subset \Omega$;

(9) $\bar{\theta}(A \cup B) = \bar{\theta}(A) \cup \bar{\theta}(B)$ *for all* $A, B \subset \Omega$;

(10) $\overline{\underline{\theta}(\bar{A})} = \bar{\theta}(A)$ *for all* $A \subset \Theta$.

§3. Is There an Ultimate Refinement?

In the past, many students of probable reasoning have sought to establish a fixed framework for their speculations by postulating the existence of an ultimately detailed set of "possible states of nature" – a frame of discernment so fine that it encompasses all possible distinctions and admits of no further refinement.* Such an ultimate refinement would offer both conceptual and mathematical advantages to a theory of probable reasoning. The possibility of working in such a frame would lighten the conceptual burden of the theory by permitting it to leave aside the problem of deciding when an instance of probable reasoning is based on a fine enough analysis of the evidence. And the existence of such a frame would simplify the theory's mathematical language by permitting all frames to be described as different partitions of the same set. But in spite of these advantages, we must reject the postulation of such an ultimate refinement.

This rejection is compelled by the purely epistemic nature of the role played by the frame of discernment. In our theory, one's frame of discernment is not a set of "states of nature" that are objectively possible in

*Compare the discussion on pp. 8-9 of Savage's *Foundations of Statistics*. Advocates of the Bayesian theory usually assume the existence of an ultimate refinement, and though they sometimes excuse this assumption as a harmless mathematical simplification, it appears to be essential to their theory. It is certainly essential to the Bayesian claim that any new evidence can be assimilated by conditioning one's global Bayesian belief function.

the sense that they are allowed by some physical law. Nor is it a set of situations that one might recognize as distinct possibilities if one knew more than one does. It is a set of possibilities that one does recognize on the basis of knowledge that one does have – or at least on the basis of distinctions that one actually draws and assumptions that one actually makes. It cannot embody concepts and distinctions that one has never heard of. Yet, as we see after a moment's reflection, each "possible state of nature" in an ultimate refinement would have to be a complete account of the history of the universe, the very writing of which would require knowledge surpassing the collective experience of mankind.

In practice, we always begin with a relatively coarse frame of discernment and then refine it as thought and evidence require, often producing distinctions that were outside the scope of our attention when we began. Hence a realistic theory of evidence will deal with frames that do not even encompass all the knowledge we do have and will explicitly allow for their refinement.

Actually, the formal nature of the frame of discernment is sufficient in itself to ensure that the frame will embody only a small part of our knowledge and experience. For alongside the exact and formal distinctions and assumptions proclaimed by the most comprehensive of our frames we will always have a much larger store of vaguer and less certain knowledge, much of it as uncertain in its expression as in its accuracy. This store of diffuse knowledge will hardly admit of a mathematical representation, but it cannot be ignored by a theory of evidence. For it will include all the impressions and hunches from which the frame is forged. And even more importantly, it will include the experience and evidence that we will bring to bear on the frame.

This last remark deserves particular emphasis, for it touches on a basic reason for avoiding, even as a simplifying idealization, the notion of an ultimate refinement. In our theory, the body of evidence that is assessed to obtain degrees of support over a frame Θ is always something that lies outside of Θ. But in any theory that claims an ultimate refinement Ω^{∞}, a body of evidence will inevitably be cast as just another *proposition* – i.e., as a subset of Ω^{∞} itself. After all, since Ω^{∞} claims to provide a means of precise expression for all knowledge and experience, it can hardly be prevented from expressing the experience that happens to constitute the evidence at issue. And such an insistence on casting all evidence in the form of propositions within the frame of discernment will lead to the debilitating conclusion that evidence cannot be assessed unless one begins with degrees of belief based on no evidence at all. We perceived this fact in §8 of Chapter 1 when we were studying the Bayesian theory, and it should be even clearer to us now. For if the evidence corresponds to a subset A of the frame of discernment, then its only effect will be to determine a simple support function focusing all one's belief on A. And this simple support function will not occasion any positive degrees of support for propositions not implied by A unless it is combined with some other belief function – one not based on the evidence.

For us, then, the frame of discernment must always remain restricted in scope. We will permit it to be indefinitely refined, but we will insist that it always fall short of formalizing all our knowledge, and especially that it fall short of formalizing the evidence that bears on it.

§4. Families of Compatible Frames

Since a frame of discernment is susceptible of indefinite refinement, there will be occasions when our attention should be directed not to that frame in particular but to the whole family of frames consisting of its refinements and the coarsenings of its refinements. The mathematical structure of such a *family of compatible frames* is delineated by the following definition.

DEFINITION. Suppose $\mathcal{F}$ is a non-empty collection of finite non-empty sets, no pair of which has any common elements. And suppose $\mathcal{R}$ is a non-empty collection of mappings, each mapping $\omega \in \mathcal{R}$ being of the form $\omega : 2^{\Theta_1} \to 2^{\Theta_2}$ for some Θ_1 and Θ_2 in $\mathcal{F}$ and qualifying as a refining by the definition of §1. Then $\mathcal{F}$ is a *family of compatible frames of discernment* with refinings $\mathcal{R}$ provided that $\mathcal{F}$ and $\mathcal{R}$ satisfy the following requirements:

(1) *Composition of Refinings*: If $\omega_1 : 2^{\Theta_1} \to 2^{\Theta_2}$ and $\omega_2 : 2^{\Theta_2} \to 2^{\Theta_3}$ are in $\mathcal{R}$, then $\omega_2 \circ \omega_1$ is in $\mathcal{R}$.

(2) *Identity of Coarsenings*: Suppose that $\omega_1 : 2^{\Theta_1} \to 2^{\Omega}$ and $\omega_2 : 2^{\Theta_2} \to 2^{\Omega}$ are elements of $\mathcal{R}$, and suppose that for each $\theta_1 \in \Theta_1$ there exists $\theta_2 \in \Theta_2$ such that $\omega_1(\{\theta_1\}) = \omega_2(\{\theta_2\})$. Then $\Theta_1 = \Theta_2$ and $\omega_1 = \omega_2$.

(3) *Identity of Refinings*: If $\omega_1 : 2^{\Theta} \to 2^{\Omega}$ and $\omega_2 : 2^{\Theta} \to 2^{\Omega}$ are elements of $\mathcal{R}$, then $\omega_1 = \omega_2$.

(4) *Existence of Coarsenings*: Suppose $\Omega \in \mathcal{F}$, and suppose $A_1, \cdots, A_n$ is a disjoint partition of Ω. Then there is a coarsening in $\mathcal{F}$ corresponding to this disjoint partition. That is to say, there exists a refining $\omega : 2^{\Theta} \to 2^{\Omega}$ in $\mathcal{R}$ such that for each i, $i = 1, \cdots, n$, there exists $\theta \in \Theta$ with $\omega(\{\theta\}) = A_i$.

(5) *Existence of Refinings*: Suppose $\Theta \in \mathcal{F}$, $\theta \in \Theta$, and n is a positive integer. Then there exists a refining $\omega : 2^{\Theta} \to 2^{\Omega}$ in $\mathcal{R}$ such that $\omega(\{\theta\})$ has n elements.

(6) *Existence of Common Refinements*: Every pair of elements in $\mathcal{F}$ has a common refinement in $\mathcal{F}$. That is to say, if $\Theta_1, \Theta_2 \in \mathcal{F}$, then there exist refinings $\omega_1 : 2^{\Theta_1} \to 2^{\Omega_1}$ and $\omega_2 : 2^{\Theta_2} \to 2^{\Omega_2}$ in $\mathcal{R}$ such that $\Omega_1 = \Omega_2$.

Whenever we discuss a frame of discernment in the balance of this essay, it is to be understood that the frame can be placed in a family of frames satisfying this definition. And when we discuss refinements or coarsenings, or "compatible frames" in general, it is to be understood that all the frames in question can be placed together in such a family.

Condition (5) of the definition is designed to express the "open-endedness" of the family $\mathcal{F}$ – i.e., the susceptibility of each of its frames to indefinite refinement. In fact, the condition is stated so as to require that this indefinite refinement already have been carried out within the family $\mathcal{F}$. Hence it is to be understood that we really never deal with more than an incomplete family of compatible frames; the notion of a whole family of compatible frames is to be evoked only as a potentiality. This potential character of a compatible family should not be forgotten, but it will not keep such families from being useful in our theory. In Chapter 8, for example, we will use the notion of a compatible family to express conditions under which a given frame is sufficiently fine to permit accurate combination of evidence, and these conditions will be intuitively meaningful even though the impossibility of having a complete family in view will mean that they can never enable us to be certain that our frame is sufficiently fine.

Using condition (6) of the definition, one may deduce the existence of a common refinement for any finite collection of compatible frames of discernment. Such a collection will in fact have many common refinements, but one of these will be distinguished by its simplicity.

THEOREM 6.4. *Suppose* $\Theta_1,\cdots,\Theta_n$ *are elements of a family* $\mathcal{F}$ *of compatible frames of discernment. Then there exists a unique element* $\Theta \in \mathcal{F}$ *such that*

(1) *for each* i, *there exists a refining* $\omega_i : 2^{\Theta_i} \to 2^{\Theta}$, *and*

(2) *for each* $\theta \in \Theta$, *there exist elements* $\theta_i \in \Theta_i$, *for* $i = 1,\cdots,n$, *such that*

$$\{\theta\} = \omega_1(\{\theta_1\}) \cap \cdots \cap \omega_n(\{\theta_n\}) .$$

(Notice that the elements θ_i are not necessarily unique; cf. Theorem 6.11 in §6 below.) The unique frame Θ specified by Theorem 6.4 is denoted $\Theta_1 \otimes \cdots \otimes \Theta_n$; it is called the *minimal refinement* of $\Theta_1,\cdots,\Theta_n$. This name is explained by Theorem 6.5.

THEOREM 6.5. *If* Ω *is a common refinement of* $\Theta_1,\cdots,\Theta_n$, *then* $\Theta_1 \otimes \cdots \otimes \Theta_n$ *is a coarsening of* Ω. *Furthermore* $\Theta_1 \otimes \cdots \otimes \Theta_n$ *is the only common refinement of* $\Theta_1,\cdots,\Theta_n$ *that has the property of being a coarsening of every other common refinement.*

As I remarked in §1 above, a proposition that is discerned by one frame will also be discerned by any refinement of that frame. Indeed, if $\omega : 2^\Theta \to 2^\Omega$ is a refining and A is a subset of Θ, then both A and $\omega(A)$ will represent the same proposition. It follows that a proposition discerned by a given frame in a compatible family will also be discerned by many of the other frames in that family. But how can we tell in general whether a subset A_1 of one frame Θ_1 represents the same proposition as a subset A_2 of a compatible frame Θ_2?

The simplest way is to refer to the minimal refinement $\Theta_1 \otimes \Theta_2$ and to the refinings $\omega_1 : 2^{\Theta_1} \to 2^{\Theta_1 \otimes \Theta_2}$ and $\omega_2 : 2^{\Theta_2} \to 2^{\Theta_1 \otimes \Theta_2}$. If $\omega_1(A_1) = \omega_2(A_2)$, then A_1 and A_2 represent the same proposition; if $\omega_1(A_1) \neq \omega_2(A_2)$, then A_1 and A_2 represent different propositions. Or, if we prefer, we may refer to any other common refinement Ω:

THEOREM 6.6. *Suppose* Ω *is a common refinement of* Θ_1 *and* Θ_2, *and consider the refinings* $\omega_1 : 2^{\Theta_1} \to 2^{\Theta_1 \otimes \Theta_2}$, $\omega'_1 : 2^{\Theta_1} \to 2^\Omega$, $\omega_2 : 2^{\Theta_2} \to 2^{\Theta_1 \otimes \Theta_2}$ *and* $\omega'_2 : 2^{\Theta_2} \to 2^\Omega$. *If* $A_1 \subset \Theta_1$ *and* $A_2 \subset \Theta_2$, *then* $\omega_1(A_1) = \omega_2(A_2)$ *if and only if* $\omega'_1(A_1) = \omega'_2(A_2)$.

According to §1 of Chapter 2, relations between a pair of propositions are supposed to be reflected by relations between the subsets that represent them. Theorem 6.7 assures us that these relations are the same for all the frames that discern the propositions.

THEOREM 6.7. *Suppose* Θ_1 *and* Θ_2 *are compatible frames, suppose* $A_1 \subset \Theta_1$ *represents the same proposition as* $A_2 \subset \Theta_2$, *and suppose* $B_1 \subset \Theta_1$ *represents the same proposition as* $B_2 \subset \Theta_2$. *Then*

(1) $A_1 \cap B_1$ *represents the same proposition as* $A_2 \cap B_2$;
(2) $A_1 \cup B_1$ *represents the same proposition as* $A_2 \cup B_2$;
(3) $A_1 \subset B_1$ *if and only if* $A_2 \subset B_2$;
(4) $A_1 = \bar{B}_1$ *if and only if* $A_2 = \bar{B}_2$;
(5) $A_1 = \emptyset$ *if and only if* $A_2 = \emptyset$;
(6) $A_1 = \Theta_1$ *if and only if* $A_2 = \Theta_2$.

I have been using the word "proposition" as an undefined term, but it should be obvious by now that a "proposition" is simply the intuitive content of the assertion that one of a set of possibilities obtains. Of course, that intuitive content depends on the whole frame and becomes richer, in a sense, as the frame is refined; hence the full meaning of a proposition discerned by a frame depends on the frame's whole compatible family and may include certain not yet realized potentialities.*

§5. Consistent Belief Functions

If Θ_1 and Θ_2 are compatible frames of discernment, then two belief functions $\mathrm{Bel}_1 : 2^{\Theta_1} \to [0,1]$ and $\mathrm{Bel}_2 : 2^{\Theta_2} \to [0,1]$ can represent the same opinion only if they agree on those propositions that are discerned by both Θ_1 and Θ_2. If this condition is met – i.e., if $\mathrm{Bel}_1(A_1) = \mathrm{Bel}_2(A_2)$ whenever $A_1 \subset \Theta_1$ and $A_2 \subset \Theta_2$ and $\omega_1(A_1) = \omega_2(A_2)$, where $\omega_1 : 2^{\Theta_1} \to 2^{\Theta_1 \otimes \Theta_2}$ and $\omega_2 : 2^{\Theta_2} \to 2^{\Theta_1 \otimes \Theta_2}$ are the refinings to the minimal refinement – then Bel_1 and Bel_2 are said to be *consistent*.

Notice that if $\omega : 2^{\Theta} \to 2^{\Omega}$ is a refining, then $\mathrm{Bel} : 2^{\Omega} \to [0,1]$ and $\mathrm{Bel}_0 : 2^{\Theta} \to [0,1]$ are consistent only if Bel_0 is given by

---

*One should *not* suppose that there is a formal language in the background of our discussion and that a proposition is always the intuitive content of a statement in that language. Such a language would amount to the same thing as an ultimate refinement.

$$\mathrm{Bel}_0(A) = \mathrm{Bel}(\omega(A)) \tag{6.2}$$

for all $A \in 2^\Theta$. If Bel_0 is indeed given by (6.2), I will call it the *restriction* of Bel to Θ and sometimes denote it by $\mathrm{Bel}|2^\Theta$. The name and the notation reflect, of course, the fact that Bel_0 does appear as a restriction of the function Bel when 2^Θ is thought of as a subset of 2^Ω.

The following theorem reassures us that the restriction of a belief function is itself always a belief function.

THEOREM 6.8. *Suppose* $\omega : 2^\Theta \to 2^\Omega$ *is a refining and* $\mathrm{Bel} : 2^\Omega \to [0,1]$ *is a belief function. Then the function* $\mathrm{Bel}_0 : 2^\Theta \to [0,1]$ *given by (6.2) is a belief function.*

This means that no restriction is placed on a belief function by the requirement that there should exist belief functions on coarsenings that are compatible with it.

The following theorem tells how to calculate the basic probability numbers and the commonality numbers for the restriction of a belief function.

THEOREM 6.9. *Suppose* $\omega : 2^\Theta \to 2^\Omega$ *is a refinement and* $\mathrm{Bel} : 2^\Omega \to [0,1]$ *is a belief function. Let* $\mathcal{C}$ *and* $\mathcal{C}_0$ *denote the cores of* Bel *and* $\mathrm{Bel}|2^\Theta$, *respectively; let* m *and* Q *denote the basic probability assignment and the commonality function for* Bel; *and let* m_0 *and* Q_0 *denote the corresponding functions for* $\mathrm{Bel}|2^\Theta$. *Let* $\bar{\theta}$ *denote the outer reduction for* ω. *Then*

$$\mathcal{C}_0 = \bar{\theta}(\mathcal{C}) .$$

And

$$m_0(A) = \sum_{\substack{B \subset \Omega \\ A = \bar{\theta}(B)}} m(B) \tag{6.3}$$

and

$$Q_0(A) = \sum_{\substack{B \subset \Omega \\ A \subset \bar{\theta}(B)}} m(B)$$

for all $A \subset \Theta$. *In particular,*

$$m_0(A) \geq m(\omega(A))$$

and

$$Q_0(A) \geq Q(\omega(A))$$

for all $A \subset \Theta$.

The last sentence of this theorem may be paraphrased by saying that the basic probability number and the commonality number for a particular proposition may decrease but may not increase as the calculation is shifted from a particular frame Θ to a refinement Ω.

§6. **Independent Frames**

Two compatible frames of discernment are *independent* if no proposition discerned by one of them non-trivially implies a proposition discerned by the other. In this section we learn how to express this condition more precisely and how to generalize it to a larger number of frames.*

To tell whether a proposition discerned by a frame Θ_1 implies a proposition discerned by a compatible frame Θ_2, we must refer to a frame that discerns both propositions, and ask whether a subset corresponding to this first proposition in that frame is contained in the subset corresponding to the second. By (3) of Theorem 6.7, it does not matter which of the frames discerning both propositions we consider; for simplicity we may consider $\Theta_1 \otimes \Theta_2$. So when Θ_1 and Θ_2 are compatible, to say that the proposition corresponding to $A_1 \subset \Theta_1$ implies the proposition corresponding to $A_2 \subset \Theta_2$ is to say that $\omega_1(A_1) \subset \omega_2(A_2)$, where $\omega_1 : 2^{\Theta_1} \to 2^{\Theta_1 \otimes \Theta_2}$ and $\omega_2 : 2^{\Theta_2} \to 2^{\Theta_1 \otimes \Theta_2}$ are the refinings to the minimal refinement. Now

---

*The intuitive notion of independence that is dealt with in this section is quite standard. For a formulation in terms of subalgebras of a Boolean algebra, see Roman Sikorski's *Boolean Algebras*, pp. 39-42.

such an implication is said to be trivial if either $A_1 = \emptyset$, so that the proposition corresponding to A_1 is trivially known to be false, or $A_2 = \Theta_2$, so that the proposition corresponding to A_2 is trivially known to be true. Hence the verbal definition of the preceding paragraph translates as follows:

DEFINITION. Suppose Θ_1 and Θ_2 are compatible frames, and denote the refinings to their minimal refinement by $\omega_1 : 2^{\Theta_1} \to 2^{\Theta_1 \otimes \Theta_2}$ and $\omega_2 : 2^{\Theta_2} \to 2^{\Theta_1 \otimes \Theta_2}$. Then Θ_1 and Θ_2 are *independent* if either $A_1 = \emptyset$ or $A_2 = \Theta_2$ whenever $A_1 \subset \Theta_1$, $A_2 \subset \Theta_2$, and $\omega_1(A_1) \subset \omega_2(A_2)$.

Theorem 6.10 provides a list of equivalent conditions.

THEOREM 6.10. *Suppose Θ_1 and Θ_2 are compatible frames, and Ω is a common refinement. Let ω_1 and ω_2 denote the corresponding refinings, and let $\underline{\theta}_1$ and $\overline{\theta}_1$ denote the inner and outer reductions for ω_1. Then the following conditions are all equivalent.*

(1) *Θ_1 and Θ_2 are independent.*

(2) *If $A_1 \subset \Theta_1$, $A_2 \subset \Theta_2$, and $\omega_1(A_1) \subset \omega_2(A_2)$, then either $A_1 = \emptyset$ or $A_2 = \Theta_2$.*

(3) *If A_1 is a non-empty subset of Θ_1 and A_2 is a non-empty subset of Θ_2, then $\omega_1(A_1) \cap \omega_2(A_2) \neq \emptyset$.*

(4) *If $\theta_1 \in \Theta_1$ and $\theta_2 \in \Theta_2$, then $\omega_1(\{\theta_1\}) \cap \omega_2(\{\theta_2\}) \neq \emptyset$.*

(5) *$\underline{\theta}_1(\omega_2(A)) = \emptyset$ for any proper subset A of Θ_2.*

(6) *$\overline{\theta}_1(\omega_2(A)) = \Theta_1$ for any non-empty subset A of Θ_2.*

We may use (3) of Theorem 6.10 to generalize the notion of independence to a collection of n compatible frames of discernment.

DEFINITION. Suppose $\Theta_1, \cdots, \Theta_n$ are compatible frames, and denote the refinings to their minimal refinement by $\omega_i : 2^{\Theta_i} \to 2^{\Theta_1 \otimes \cdots \otimes \Theta_n}$ for

$i = 1, \cdots, n$. Then $\Theta_1, \cdots, \Theta_n$ are *independent* if

$$\omega_1(A_1) \cap \cdots \cap \omega_n(A_n) \neq \emptyset$$

whenever $\emptyset \neq A_i \subset \Theta_i$ for $i = 1, \cdots, n$.

THEOREM 6.11. *Suppose $\Theta_1, \cdots, \Theta_n$ are compatible frames and Ω is a common refinement. Let $\omega_1, \cdots, \omega_n$ denote the corresponding refinings. Then the following conditions are all equivalent.*

(1) *$\Theta_1, \cdots, \Theta_n$ are independent.*

(2) *If $A_i \subset \Theta_i$ for $i = 1, \cdots, n$, and*

$$\omega_1(A_1) \cap \cdots \cap \omega_{n-1}(A_{n-1}) \subset \omega_n(A_n) ,$$

then either one of the first $n-1$ A_i is empty or else $A_n = \Theta_n$.

(3) *If $\emptyset \neq A_i \subset \Theta$ for $i = 1, \cdots, n$, then $\omega_1(A_1) \cap \cdots \cap \omega_n(A_n) \neq \emptyset$.*

(4) *If $\theta_i \in \Theta$ for $i = 1, \cdots, n$, then $\omega_1(\{\theta_1\}) \cap \cdots \cap \omega_n(\{\theta_n\}) \neq \emptyset$.*

§7. Mathematical Appendix

Proof of Theorem 6.1.

(1) Suppose $A, B \subset \Theta$, and $A \neq B$. Then since the $\omega(\{\theta\})$ are non-empty and form a disjoint partition,

$$\omega(A) = \bigcup_{\theta \in A} \omega(\{\theta\}) \neq \bigcup_{\theta \in B} \omega(\{\theta\}) = \omega(B) .$$

(2) $\omega(\emptyset) = \bigcup_{\theta \in \emptyset} \omega(\{\theta\}) = \emptyset$; this is in accordance with the usual convention that the union of an empty collection is empty.

(3) $\omega(\Theta) = \bigcup_{\theta \in \Theta} \omega(\{\theta\}) = \Omega$; this is because the $\omega(\{\theta\})$ are a partition of Ω.

(4)

$$\omega(A \cup B) = \bigcup_{\theta \in A \cup B} \omega(\{\theta\}) = \Big(\bigcup_{\theta \in A} \omega(\{\theta\})\Big) \cup \Big(\bigcup_{\theta \in B} \omega(\{\theta\})\Big) = \omega(A) \cup \omega(B).$$

(5) Since the $\omega(\{\theta\})$ are all disjoint,

$$\omega(A) = \bigcup_{\theta \in A} \omega(\{\theta\}) \quad \text{and} \quad \omega(\overline{A}) = \bigcup_{\theta \in \overline{A}} \omega(\{\theta\})$$

are disjoint. And since the $\omega(\{\theta\})$ are a partition, the union of these two sets is the whole set Ω. Hence $\omega(\overline{A})$ is the complement of $\omega(A)$.

(6) Using (4) and (5),

$$\omega(A \cap B) = \omega(\overline{\overline{A} \cup \overline{B}}) = \overline{\omega(\overline{A} \cup \overline{B})} = \overline{\omega(\overline{A}) \cup \omega(\overline{B})} = \overline{\overline{\omega(A)} \cup \overline{\omega(B)}} = \omega(A) \cap \omega(B).$$

(7) and (8) both follow immediately from (6.1), together with the disjointness of the $\omega(\{\theta\})$. ■

Proof of Theorem 6.2.

(i) $\omega(B) \subset A$ if and only if $\omega(\{\theta\}) \subset A$ for all $\theta \in B$ – i.e., if and only if

$$B \subset \{\theta \in \Theta | \omega(\{\theta\}) \subset A\} = \underline{\theta}(A).$$

(ii) Since the $\omega(\{\theta\})$ form a disjoint partition of Ω, $A \subset \omega(B)$ if and only if B includes every $\theta \in \Theta$ such that $\omega(\{\theta\})$ has a non-empty intersection with A – i.e., if and only if

$$\overline{\theta}(A) = \{\theta \in \Theta | \omega(\{\theta\}) \cap A \neq \emptyset\} \subset B. \ ■$$

Proof of Theorem 6.3.

(1) is immediate, for since the $\omega(\{\theta\})$ are non-empty, $\omega(\{\theta\}) \cap A \neq \emptyset$ whenever $\omega(\{\theta\}) \subset A$.

(2) By (7) of Theorem 6.1,

$$\underline{\theta}(\omega(A)) = \{\theta \in \Theta | \omega(\{\theta\}) \subset \omega(A)\} = \{\theta \in \Theta | \{\theta\} \subset A\} = A,$$

and by (8) of Theorem 6.1,

$$\overline{\theta}(\omega(A)) = \{\theta \in \Theta | \omega(\{\theta\}) \cap \omega(A) \neq \emptyset\} = \{\theta \in \Theta | \{\theta\} \cap A \neq \emptyset\} = A.$$

(3) follows from Theorem 6.2.

(4) follows from Theorem 6.2.

(5) follows from (2) of Theorem 6.1 and (2) of this theorem.

(6) follows from (3) of Theorem 6.1 and (2) of this theorem.

(7) If $A \subset B$, then $\omega(\{\theta\}) \subset B$ whenever $\omega(\{\theta\}) \subset A$, and $\omega(\{\theta\}) \cap B \neq \emptyset$ whenever $\omega(\{\theta\}) \cap A \neq \emptyset$.

(8)
$$\begin{aligned}\underline{\theta}(A\cap B) &= \{\theta \in \Theta \mid \omega(\{\theta\}) \subset A \cap B\} \\ &= \{\theta \in \Theta \mid \omega(\{\theta\}) \subset A \text{ and } \omega(\{\theta\}) \subset B\} \\ &= \{\theta \in \Theta \mid \omega(\{\theta\}) \subset A\} \cap \{\theta \in \Theta \mid \omega(\{\theta\}) \subset B\} \\ &= \underline{\theta}(A) \cap \underline{\theta}(B).\end{aligned}$$

(9)
$$\begin{aligned}\overline{\theta}(A\cup B) &= \{\theta \in \Theta \mid \omega(\{\theta\}) \cap (A\cup B) \neq \emptyset\} \\ &= \{\theta \in \Theta \mid \omega(\{\theta\}) \cap A \neq \emptyset \text{ or } \omega(\{\theta\}) \cap B \neq \emptyset\} \\ &= \{\theta \in \Theta \mid \omega(\{\theta\}) \cap A \neq \emptyset\} \cup \{\theta \in \Theta \mid \omega(\{\theta\}) \cap B \neq \emptyset\} \\ &= \overline{\theta}(A) \cup \overline{\theta}(B).\end{aligned}$$

(10)
$$\begin{aligned}\overline{\underline{\theta}(\overline{A})} &= \{\theta \in \Theta \mid \omega(\{\theta\}) \not\subset \overline{A}\} \\ &= \{\theta \in \Theta \mid \omega(\{\theta\}) \cap A \neq \emptyset\} \\ &= \overline{\theta}(A). \ \blacksquare\end{aligned}$$

LEMMA 6.1. *Any finite collection of compatible frames of discernment has a common refinement.*

Proof of Lemma 6.1. This lemma can be proven by induction on the number n of frames in the collection, provided we use conditions (1) and (6) in the definition of a family of compatible frames.

When $n = 2$, the lemma is simply a restatement of condition (6). Suppose the lemma holds when $n = k$, and suppose $n = k+1$. Then the frames in the collection may be enumerated as $\Theta_1, \cdots, \Theta_{k+1}$, and it may be assumed that the first k have a common refinement Ω_0, with refinings $\omega_i : 2^{\Theta_i} \to 2^{\Omega_0}$ for $i = 1, \cdots, k$. But by (6) there is a common refinement Ω for Ω_0 and Θ_{k+1}, with refinings $\omega_0 : 2^{\Omega_0} \to 2^{\Omega}$ and $\omega_{k+1} : 2^{\Theta_{k+1}} \to 2^{\Omega}$. And by (2), $(\omega_0 \circ \omega_i) : 2^{\Theta_i} \to 2^{\Omega}$ is a refining for each i, $i = 1, \cdots, k$. So Ω is a common refinement for $\Theta_1, \cdots, \Theta_{k+1}$. $\blacksquare$

LEMMA 6.2. *Suppose* $\omega_1 : 2^{\Theta_1} \to 2^\Omega$, $\omega_2 : 2^{\Theta_2} \to 2^\Omega, \cdots, \omega_n : 2^{\Theta_n} \to 2^\Omega$ *are all refinings. Then the non-empty sets of the form*

$$\omega_1(\{\theta_1\}) \cap \cdots \cap \omega_n(\{\theta_n\}),$$

where $\theta_i \in \Theta_i$ *for each* i, *constitute a disjoint partition of* Ω.

Proof of Lemma 6.2. If $\theta_i \neq \theta'_i$, then $\omega_i(\{\theta_i\}) \cap \omega_i(\{\theta'_i\}) = \emptyset$. It follows that two sets

$$\omega_1(\{\theta_1\}) \cap \cdots \cap \omega_n(\{\theta_n\})$$

and

$$\omega_1(\{\theta'_1\}) \cap \cdots \cap \omega_n(\{\theta'_n\})$$

are either identical or disjoint.

Now for each i, $i = 1, \cdots, n$, the $\omega_i(\{\theta_i\})$ constitute a disjoint partition of Ω. Hence for every element $\theta \in \Omega$, and each i, there exists an element $\theta_i \in \Theta_i$ such that $\theta \in \omega_i(\{\theta_i\})$. Choosing such a θ_i for each i yields

$$\theta \in \omega_1(\{\theta_1\}) \cap \cdots \cap \omega_n(\{\theta_n\}).$$

Since this can be done for every $\theta \in \Omega$, the union of the sets of this form is all of Ω; they form a disjoint partition. ∎

LEMMA 6.3. *Suppose that* $\omega_1 : 2^{\Theta_1} \to 2^\Omega$ *and* $\omega_2 : 2^{\Theta_2} \to 2^\Omega$ *are refinings, and suppose that for each* $\theta \in \Theta_1$ *there exists* $A \subset \Theta_2$ *such that* $\omega_1(\{\theta\}) = \omega_2(A)$. *Then* Θ_2 *is a refinement of* Θ_1.

Proof of Lemma 6.3. Enumerate the elements of Θ_1, denoting them $\theta_1, \cdots, \theta_n$. Choose $A_1, \cdots, A_n$ such that $\omega_1(\{\theta_i\}) = \omega_2(A_i)$. If $i \neq j$, then $\{\theta_i\}$ and $\{\theta_j\}$ are disjoint; it follows by Theorem 6.1 that $\omega_1(\{\theta_i\}) = \omega_2(A_i)$ and $\omega_1(\{\theta_j\}) = \omega_2(A_j)$ are disjoint, and then that A_i and A_j are disjoint. Again using Theorem 6.1,

$$\omega_2\left(\bigcup_{i=1}^{n} A_i\right) = \bigcup_{i=1}^{n} \omega_2(A_i) = \bigcup_{i=1}^{n} \omega_1(\{\theta_i\}) = \omega_1\left(\bigcup_{i=1}^{n} \{\theta_i\}\right)$$

$$= \omega_1(\Theta_1) = \Omega ,$$

so

$$\bigcup_{i=1}^{n} A_i = \Theta_2 .$$

So $A_1,\cdots,A_n$ is a disjoint partition of Θ_2. It follows by condition (4) of the definition of a family of compatible frames that there exists a refining $\omega : 2^{\Theta} \to 2^{\Theta_2}$ such that for each i, $i = 1,\cdots,n$, there exists $\theta \epsilon \Theta$ with $\omega(\{\theta\}) = A_i$. Notice that the refining $\omega_2 \circ \omega$ maps Θ into Ω, and that for each $\theta_i \epsilon \Theta_1$, there exists $\theta \epsilon \Theta$ such that $\omega_1(\{\theta_i\}) = \omega_2(A_i) = (\omega_2 \circ \omega)(\{\theta\})$. Hence $\omega_1 = \omega_2 \circ \omega$ and $\Theta_1 = \Theta$, by condition (2) of the definition of a family of compatible frames. This means that Θ_2 is a refinement of Θ_1. ■

Proof of Theorem 6.4.

(i) Let us begin by showing that there is at most one frame satisfying (1) and (2).

Suppose, indeed, that Θ and Θ' are in $\mathcal{F}$ and both satisfy (1) and (2). That is to say, there are refinings $\omega_i : 2^{\Theta_i} \to 2^{\Theta}$ and $\omega'_i : 2^{\Theta_i} \to 2^{\Theta'}$ for each i, $i = 1,\cdots,n$. And for each $\theta \epsilon \Theta$, there exist elements $\theta_i \epsilon \Theta_i$, for $i = 1,\cdots,n$, such that

$$\{\theta\} = \omega_1(\{\theta_1\}) \cap \cdots \cap \omega_n(\{\theta_n\}) ,$$

while for each $\theta' \epsilon \Theta'$, there exist elements $\theta'_i \epsilon \Theta_i$, for $i = 1,\cdots,n$, such that

$$\{\theta'\} = \omega'_1(\{\theta'_1\}) \cap \cdots \cap \omega'_n(\{\theta'_n\}) .$$

Notice that by Lemma 6.2 the non-empty subsets of Θ' of the form

$$\omega'_1(\{\theta_1\}) \cap \cdots \cap \omega'_n(\{\theta_n\}) ,$$

where $\theta_i \epsilon \Theta_i$, constitute a disjoint partition of Θ'. It follows that every non-empty subset of this form is equal to some singleton $\{\theta'\} \subset \Theta'$.

Let Ω denote a common refinement of Θ and Θ', and let $\omega : 2^{\Theta} \to 2^{\Omega}$ and $\omega' : 2^{\Theta'} \to 2^{\Omega}$ denote the refinings. Notice that for each i, both $\omega \circ \omega_i$ and $\omega' \circ \omega'_i$ map 2^{Θ_i} into 2^{Ω}. It follows by condition (3) of our definition that $\omega \circ \omega_i = \omega' \circ \omega'_i$. Suppose $\theta \epsilon \Theta$. Then there exist elements $\theta_i \epsilon \Theta_i$, $i = 1, \cdots, n$, such that

$$\{\theta\} = \omega_1(\{\theta_1\}) \cap \cdots \cap \omega_n(\{\theta_n\}) ,$$

or

$$\begin{aligned} \omega(\{\theta\}) &= \omega(\omega_1(\{\theta_1\}) \cap \cdots \cap \omega_n(\{\theta_n\})) \\ &= (\omega \circ \omega_1)(\{\theta_1\}) \cap \cdots \cap (\omega \circ \omega_n)(\{\theta_n\}) \\ &= (\omega' \circ \omega'_1)(\{\theta_1\}) \cap \cdots \cap (\omega' \circ \omega'_n)(\{\theta_n\}) \\ &= \omega'(\omega'_1(\{\theta_1\}) \cap \cdots \cap \omega'_n(\{\theta_n\})) , \end{aligned}$$

and by the preceding paragraph, this is equal to $\omega'(\{\theta'\})$ for some $\theta' \epsilon \Theta'$. Since $\theta \epsilon \Theta$ was arbitrary, it follows from condition (2) of our definition that $\Theta = \Theta'$.

(ii) Let us now construct a frame Θ satisfying (1) and (2).

By Lemma 6.1, the frames $\Theta_1, \cdots, \Theta_n$ have a common refinement Ω. For each i, $i = 1, \cdots, n$, let ω'_i denote the refining that maps 2^{Θ_i} into 2^{Ω}. Then by Lemma 6.2, a disjoint partition of Ω is constituted by the non-empty sets of the form

$$\omega'_1(\{\theta_1\}) \cap \cdots \cap \omega'_n(\{\theta_n\}) ,$$

where $\theta_i \epsilon \Theta_i$ for each i. By condition (4) of the definition of a compatible family, there exists a coarsening Θ of Ω, with refining $\omega : 2^{\Theta} \to 2^{\Omega}$, such that this disjoint partition coincides with the disjoint partition formed by the $\omega(\{\theta\})$.

Now let us show that Θ satisfies (1). Let us prove, for example, that Θ is a refinement of Θ_1. Fix $\theta_1 \epsilon \Theta_1$. Let θ_i denote an arbitrary element of Θ_i, $i = 2,\cdots,n$. Then by the construction of Θ, either

$$\omega'_1(\{\theta_1\}) \cap \cdots \cap \omega'_n(\{\theta_n\}) = \emptyset$$

or else there exists an element of Θ, which we may denote as $\theta(\theta_2,\cdots,\theta_n)$, such that

$$\omega(\{\theta(\theta_2,\cdots,\theta_n)\}) = \omega'_1(\{\theta_1\}) \cap \cdots \cap \omega'_n(\{\theta_n\}) .$$

Set

$$A = \{\theta \in \Theta | \theta = \theta(\theta_2,\cdots,\theta_n) \text{ for some } \theta_2 \in \Theta_2,\cdots,\theta_n \in \Theta_n\} .$$

Then

$$\begin{aligned} \omega(A) &= \cup\{\omega(\{\theta\}) | \theta = \theta(\theta_2,\cdots,\theta_n) \text{ for some } \theta_2 \in \Theta_2,\cdots,\theta_n \in \Theta_n\} \\ &= \cup\{\omega'_1(\{\theta_1\}) \cap \cdots \cap \omega'_n(\{\theta_n\}) | \theta_2 \in \Theta_2,\cdots,\theta_n \in \Theta_n\} \\ &= \omega'_1(\{\theta_1\}) \cap (\cup\{\omega'_2(\{\theta_2\}) \cap \cdots \cap \omega'_n(\{\theta_n\}) | \theta_2 \in \Theta_2,\cdots,\theta_n \in \Theta_n\}) \\ &= \omega'_1(\{\theta_1\}) . \end{aligned}$$

Since θ_1 was an arbitrary element of Θ_1, we have proven that for every $\theta_1 \in \Theta_1$ there exists $A \subset \Theta$ such that $\omega'_1(\{\theta_1\}) = \omega(A)$. It follows by Lemma 6.3 that Θ is a refinement of Θ_1.

Similarly, one finds that Θ is a refinement of each of the Θ_i. We may denote the refinings in question by $\omega_i : 2^{\Theta_i} \to 2^{\Theta}$.

Notice that for each i, $i = 1,\cdots,n$, both ω'_i and $\omega \circ \omega_i$ map 2^{Θ_i} into 2^{Ω}. Hence $\omega'_i = \omega \circ \omega_i$, by condition (3) of the definition of a compatible family. By the construction of Θ, then, there exists, for each element $\theta \in \Theta$, a collection of elements $\theta_i \in \Theta_i$, $i = 1,\cdots,n$, such that

$$\begin{aligned} \omega(\{\theta\}) &= \omega'_1(\{\theta_1\}) \cap \cdots \cap \omega'_n(\{\theta_n\}) \\ &= \omega(\omega_1(\{\theta_1\})) \cap \cdots \cap \omega(\omega_n(\{\theta_n\})) \\ &= \omega(\omega_1(\{\theta_1\}) \cap \cdots \cap \omega_n(\{\theta_n\})) . \end{aligned}$$

And since ω is one-to-one, this means that

$$\{\theta\} = \omega_1(\{\theta_1\}) \cap \cdots \cap \omega_n(\{\theta_n\}) .$$

So Θ satisfies (2). ■

LEMMA 6.4. *Suppose* Θ *is a frame in a family* $\mathcal{F}$ *of compatible frames with refinings* $\mathcal{R}$. *Then the identity mapping from* 2^Θ *to* 2^Θ *is in* $\mathcal{R}$.

Proof of Lemma 6.4. Suppose there are n elements in Θ, and denote them by $\theta_1,\cdots,\theta_n$. Then $\{\theta_1\},\cdots,\{\theta_n\}$ constitute a disjoint partition of Θ, and by condition (4) of the definition of a family of compatible frames, there exists a frame Θ' in $\mathcal{F}$ and a refining $\omega':2^{\Theta'} \to 2^\Theta$ in $\mathcal{R}$ such that for each i, $i = 1,\cdots,n$, there exists $\theta'_i \in \Theta'$ satisfying $\omega'(\{\theta'_i\})=\{\theta_i\}$.

By condition (5), $\mathcal{F}$ contains a refinement Ω of Θ; let $\omega : 2^\Theta \to 2^\Omega$ denote the refining. Then both ω and $\omega \circ \omega'$ map into 2^Ω. And for every $\theta \in \Theta$, there exists $\theta' \in \Theta'$ such that $\omega(\{\theta\}) = \omega(\omega'(\{\theta'\})) = (\omega \circ \omega')(\{\theta'\})$. So by condition (2), $\Theta = \Theta'$ and $\omega = \omega \circ \omega'$.

Since ω is one-to-one, the relation $\omega(\{\theta\}) = \omega(\omega'(\{\theta\}))$ implies $\omega'(\{\theta\}) = \{\theta\}$ for all $\theta \in \Theta$. It then follows by (6.1) that $\omega'(A) = A$ for all $A \subset \Theta$; ω' is the identity. ■

Proof of Theorem 6.5. Let Θ denote $\Theta_1 \otimes \cdots \otimes \Theta_n$, and let $\omega_i : 2^{\Theta_i} \to 2^\Theta$ and $\omega'_i : 2^{\Theta_i} \to 2^\Omega$ denote the refinings, $i = 1,\cdots,n$.

The frames Θ and Ω will have a common refinement, say Ω_0. Denote the refinings by $\omega : 2^\Theta \to 2^{\Omega_0}$ and $\omega' : 2^\Omega \to 2^{\Omega_0}$. Now for each i, $i = 1,\cdots,n$, both $\omega \circ \omega_i$ and $\omega' \circ \omega'_i$ map 2^{Θ_i} into 2^{Ω_0}. Hence $\omega \circ \omega_i = \omega' \circ \omega'_i$, by condition (3) of the definition of a compatible family. Now consider any element $\theta \in \Theta$. By the definition of $\Theta_1 \otimes \cdots \otimes \Theta_n$, there exist elements $\theta_1 \in \Theta_1,\cdots,\theta_n \in \Theta_n$ such that

$$\{\theta\} = \omega_1(\{\theta_1\}) \cap \cdots \cap \omega_n(\{\theta_n\}) ,$$

so that

$$\begin{aligned}\omega(\{\theta\}) &= \omega(\omega_1(\{\theta_1\}))\cap\cdots\cap\omega(\omega_n(\{\theta_n\}))\\ &= \omega'(\omega'_1(\{\theta_1\})\cap\cdots\cap\omega'_n(\{\theta_n\}))\,.\end{aligned}$$

It follows by Lemma 6.3 that Ω is a refinement of Θ.

Now suppose Ω' is another common refinement of $\Theta_1,\cdots,\Theta_n$ that has the property of being a coarsening of every other common refinement of $\Theta_1,\cdots,\Theta_n$. Then Θ is a refinement of Ω', and Ω' is a refinement of Θ. Denote the refinings by $\omega_0 : 2^\Theta \to 2^{\Omega'}$ and $\omega^0 : 2^{\Omega'} \to 2^\Theta$. Then $\omega^0\circ\omega_0$ maps 2^Θ into 2^Θ; by Lemma 6.4 and condition (3), $\omega^0\circ\omega_0$ must be the identity. Hence

$$\{\theta\} = (\omega^0\circ\omega_0)(\{\theta\}) = \omega^0(\omega_0(\{\theta\}))$$

for each $\theta\in\Theta$. It follows that $\omega_0(\{\theta\})$ must be a singleton subset of Ω' for each $\theta\in\Theta$. So for every $\theta\in\Theta$, there exists $\theta'\in\Theta$ such that $(\omega^0\circ\omega_0)(\{\theta\}) = \omega^0(\{\theta'\})$. It follows by condition (2) that $\omega^0\circ\omega_0 = \omega^0$ and $\Theta = \Omega'$. ■

Proof of Theorem 6.6. By Theorem 6.5, Ω is a refinement of $\Theta_1\otimes\Theta_2$; if we denote the refining by $\omega : 2^{\Theta_1\otimes\Theta_2} \to 2^\Omega$, then $\omega'_1 = \omega\circ\omega_1$ and $\omega'_2 = \omega\circ\omega_2$, by condition (3) of the definition of a compatible family. Thus $\omega'_1(A_1) = \omega(\omega_1(A_1))$ and $\omega'_2(A_2) = \omega(\omega_2(A_2))$, and the theorem follows because ω is one-to-one. ■

Proof of Theorem 6.7. Let $\omega_1 : 2^{\Theta_1} \to 2^{\Theta_1\otimes\Theta_2}$ and $\omega_2 : 2^{\Theta_2} \to 2^{\Theta_1\otimes\Theta_2}$ denote the refinings to the minimal refinement. Then the hypothesis of the theorem is that $\omega_1(A_1) = \omega_2(A_2)$ and $\omega_1(B_1) = \omega_2(B_2)$. And the various assertions of the theorem follow from the various properties of refinings, listed in Theorem 6.1. For example,

$$\begin{aligned}\omega_1(A_1\cap B_1) &= \omega_1(A_1)\cap\omega_1(B_1)\\ &= \omega_2(A_2)\cap\omega_2(B_2)\\ &= \omega_2(A_2\cap B_2)\,,\end{aligned}$$

so $A_1\cap B_1$ represents the same proposition as $A_2\cap B_2$. ■

Proof of Theorem 6.8. We need only establish the three conditions of Theorem 2.1.

(1) $\mathrm{Bel}_0(\emptyset) = \mathrm{Bel}(\omega(\emptyset)) = \mathrm{Bel}(\emptyset) = 0$.

(2) $\mathrm{Bel}_0(\Theta) = \mathrm{Bel}(\omega(\Theta)) = \mathrm{Bel}(\Omega) = 1$.

(3) For every positive integer n and every collection $A_1, \cdots, A_n$ of subsets of Θ,

$$\begin{aligned}
\mathrm{Bel}_0(A_1 \cup \cdots \cup A_n) &= \mathrm{Bel}(\omega(A_1 \cup \cdots \cup A_n)) \\
&= \mathrm{Bel}(\omega(A_1) \cup \cdots \cup \omega(A_n)) \\
&\geq \sum_{\substack{I \subset \{1,\ldots,n\} \\ I \neq \emptyset}} (-1)^{|I|+1} \mathrm{Bel}\Big(\bigcap_{i \in I} \omega(A_i)\Big) \\
&= \sum_{\substack{I \subset \{1,\cdots,n\} \\ I \neq \emptyset}} (-1)^{|I|+1} \mathrm{Bel}\Big(\omega\Big(\bigcap_{i \in I} A_i\Big)\Big) \\
&= \sum_{\substack{I \subset \{1,\cdots,n\} \\ I \neq \emptyset}} (-1)^{|I|+1} \mathrm{Bel}_0\Big(\bigcap_{i \in I} A_i\Big) . \blacksquare
\end{aligned}$$

Proof of Theorem 6.9. Let us assume that m_0 is given by (6.3), and then prove that it is the basic probability assignment for Bel_0. Using Theorem 6.2, we find that

$$\begin{aligned}
\sum_{B \subset A} m_0(B) &= \sum_{B \subset A} \sum_{\substack{C \subset \Omega \\ B = \bar{\theta}(C)}} m(C) \\
&= \sum_{\substack{C \subset \Omega \\ \bar{\theta}(C) \subset A}} m(C) \\
&= \sum_{C \subset \omega(A)} m(C) \\
&= \mathrm{Bel}(\omega(A)) \\
&= \mathrm{Bel}_0(A) .
\end{aligned}$$

So by Theorem 2.2, m_0 is indeed the basic probability assignment for Bel_0.

Hence

$$\begin{aligned}\mathcal{C}_0 &= \cup\{A\subset\Theta \mid m_0(A) > 0\} \\ &= \cup\{A\subset\Theta \mid \text{ there exists } B\subset\Omega \text{ such that } A=\bar{\theta}(B) \text{ and } m(B)>0\} \\ &= \cup\{\bar{\theta}(B) \mid B\subset\Omega \text{ and } m(B)>0\} \\ &= \bar{\theta}(\cup\{B\subset\Omega \mid m(B)>0\}) \\ &= \bar{\theta}(\mathcal{C}).\end{aligned}$$

And

$$Q_0(A) = \sum_{\substack{B\subset\Theta \\ A\subset B}} m_0(B) = \sum_{\substack{B\subset\Theta \\ A\subset B}} \sum_{\substack{C\subset\Omega \\ B=\bar{\theta}(C)}} m(C)$$

$$= \sum_{\substack{C\subset\Omega \\ A\subset\bar{\theta}(C)}} m(C)\ .$$

Now $\bar{\theta}(\omega(A)) = A$ for each $A\subset\Theta$, and hence $A\subset\bar{\theta}(B)$ whenever $\omega(A)\subset B$. So

$$m_0(A) = \sum_{\substack{B\subset\Omega \\ A=\bar{\theta}(B)}} m(B) \geq m(\omega(A))\ ,$$

and

$$Q_0(A) = \sum_{\substack{B\subset\Omega \\ A\subset\bar{\theta}(B)}} m(B) \geq \sum_{\substack{B\subset\Omega \\ \omega(A)\subset B}} m(B) = Q(\omega(A))\ . \ \blacksquare$$

Proof of Theorem 6.10.

(i) (1) is equivalent to (2). This follows immediately from (3) of Theorem 6.7.

(ii) (2) is equivalent to (3). Indeed, $\omega_1(A_1)\cap\omega_2(A_2)\neq\emptyset$ for all non-empty A_1 and A_2 if and only if

$$\omega_1(A_1) \not\subset \overline{\omega_2(A_2)} = \omega_2(\bar{A}_2)$$

for all non-empty A_1 and A_2 – i.e., if and only if $\omega_1(A_1) \subset \omega_2(\bar{A}_2)$ can hold only when A_1 is empty or A_2 is empty.

(iii) (3) is equivalent to (4). Indeed, (4) may be obtained from (3) by substituting $\{\theta_1\}$ for A_1 and $\{\theta_2\}$ for A_2. And since

$$\omega_1(A_1) \cap \omega_2(A_2) = \bigcup_{\theta_1 \epsilon A_1,\, \theta_2 \epsilon A_2} \omega_1(\{\theta_1\}) \cap \omega_2(\{\theta_2\}) ,$$

(3) follows from (4) as well.

(iv) (4) is equivalent to (6). For $\bar{\theta}_1(\omega_2(A)) = \Theta_1$ for all non-empty $A \subset \Theta_2$ if and only if it holds for all singletons $\{\theta_2\} \subset \Theta_2$, and

$$\bar{\theta}_1(\omega_2(\{\theta_2\})) = \{\theta_1 \epsilon \Theta_1 | \omega_1(\{\theta_1\}) \cap \omega_2(\{\theta_2\}) \neq \emptyset\} .$$

(v) (5) is equivalent to (6). For $\emptyset = \theta_1(\omega_2(A))$ if and only if

$$\Theta_1 = \overline{\underline{\theta}_1(\omega_2(A))} = \bar{\theta}_1(\overline{\omega_2(A)}) = \bar{\theta}_1(\omega_2(\bar{A})) . \blacksquare$$

Proof of Theorem 6.11. The equivalence of (1) and (3) follows from (1) of Theorem 6.7. The equivalence of (2), (3) and (4) can be established by arguments exactly analogous to those of the preceding proof. ■

CHAPTER 7. SUPPORT FUNCTIONS

According even to the avowed doctrines of Protagoras and Gorgias, no truth could claim any higher value than that of a plausible opinion.

CONNOP THIRLWALL (1797-1875)

Equipped with the ideas of the preceding chapter, we can now proceed to study the *support functions* – the belief functions that can be obtained from separable support functions by coarsening the frame of discernment. The support functions are central to this essay, for they seem to constitute the subclass of belief functions appropriate for the representation of evidence. They are characterized in §1 of this chapter, and the full vocabulary associated with them is adduced in §2.

In §§3-5 we study the relations among support functions on compatible frames of discernment. The theory of §3 deals with those situations in which a given refinement does not result in any finer analysis of given evidence; this theory will be used repeatedly in the following chapters. The theory of §§4 and 5 deals with two ways in which frames that are independent in the sense of the last chapter may or may not be independent with respect to the evidence; this theory will find less application in the remainder of the essay, but it stands on its own as a pleasing example of the richness of our theory.

For the most part, the theorems of §§3-5 can be immediately generalized to apply to belief functions in general. But the motivation for those theorems depends on the idea that the functions in question are based on evidence, and hence it seems best to present them as part of the theory of support functions.

§1. The Class of Support Functions

The fact that Dempster's rule of combination is applicable only to set functions that satisfy our axioms for belief functions has led us to suppose – or to hope – that the class of belief functions is sufficiently large to represent the impact of any body of evidence on any fixed frame of discernment. This supposition does not, however, commit us to the idea that all belief functions are useful for the representation of evidence; the belief functions that can arise from actual evidence might comprise only a proper subset of the class of all belief functions. And in fact, we have thus far established only a relatively small class of belief functions as clearly appropriate for the representation of evidence. These belief functions are called *separable support functions* and represent evidence analyzable into components that are homogeneous with respect to a given frame of discernment – i.e., into components each of which precisely and unambiguously supports a given proposition discerned by that frame.

Now that we are familiar with the notion of coarsening, we can perceive a larger class of belief functions that can arise from evidence. Indeed, if given evidence can be analyzed into components homogeneous with respect to a frame Ω, so that it induces a separable support function $S : 2^{\Omega} \to [0,1]$, then that evidence's impact on a coarsening Θ will be represented by the restriction $S|2^{\Theta}$, which will be a belief function even if it is not itself a separable support function.

DEFINITION.* A belief function $\text{Bel}: 2^{\Theta} \to [0,1]$ is a *support function* if there exists some refinement Ω of Θ and some separable support function $S : 2^{\Omega} \to [0,1]$ such that $\text{Bel} = S|2^{\Theta}$.

---

*The use of the term "support function" specified by this definition differs from the use I made of the term in my paper "A Theory of Statistical Evidence."

Notice that this definition allows separable support functions to qualify as support functions, so that we obtain the inclusions

$$\left\{\begin{matrix}\text{simple}\\ \text{support}\\ \text{functions}\end{matrix}\right\} \subset \left\{\begin{matrix}\text{separable}\\ \text{support}\\ \text{functions}\end{matrix}\right\} \subset \left\{\begin{matrix}\text{support}\\ \text{functions}\end{matrix}\right\} \subset \left\{\begin{matrix}\text{belief}\\ \text{functions}\end{matrix}\right\}, \quad (7.1)$$

as advertised in the introduction to Chapter 4.

We know from Chapter 4 that not all separable support functions are simple. And using the next two theorems, we may verify that the two other inclusions in (7.1) are also proper.

THEOREM 7.1.* *Suppose* Bel *is a belief function, and* $\mathcal{C}$ *is its core. Then the following conditions are all equivalent*:

(1) Bel *is a support function.*

(2) $\mathcal{C}$ *has a positive commonality number.*

(3) $\mathcal{C}$ *has a positive basic probability number.*

THEOREM 7.2. *Suppose* S *is a separable support function,* A *and* B *are two of its focal elements, and* $A \cap B \neq \emptyset$. *Then* $A \cap B$ *is a focal element of* S.

Indeed, Theorem 7.1 implies that the belief function of Example 2.1 is not a support function. And the two theorems together suggest how to construct a support function that is not separable:

EXAMPLE 7.1. *A Non-separable Support Function.* Set $\Theta = \{a,b,c\}$, and define a belief function Bel over Θ that has focal elements $\{a,b\}$, $\{a,c\}$, and Θ – say each of these three sets has a basic probability number equal to $\frac{1}{3}$. Since Θ is a focal element, Theorem 7.1 implies that Bel is a support function. But since $\{a,b\} \cap \{a,c\} = \{a\}$ is not a focal element, Theorem 7.2 implies Bel is not a separable support function. ∎

*This theorem depends crucially on the possibility of refining any frame of discernment. It would not be true if we were to assume the existence of an ultimate refinement instead of adopting the definition of a compatible family given in §4 of Chapter 6.

Those belief functions that do not qualify as support functions will be studied more closely in Chapter 9. As we will see there, such belief functions can be intuitively interpreted in terms of our theory, but that interpretation disqualifies them for the representation of actual evidence. Hence we are led to the idea that the class of support functions, as its name suggests, might itself be sufficiently broad to represent the impact of any body of evidence on any finite frame of discernment.

§2. Support, Dubiety, and Plausibility

An adequate summary of the impact of the evidence on a particular proposition A must include at least two items of information: a report on how well A is supported and a report on how well its negation $\bar{A}$ is supported. Since a proposition is *plausible* in light of the evidence to the extent that that evidence does not support its negation, these two items of information can be conveyed by reporting the proposition's degree of support, S(A), together with its *degree of plausibility*,

$$Pl(A) = 1 - S(\bar{A}) . \tag{7.2}$$

Like the degree of support, the degree of plausibility will be a number between zero and one; it will be zero when the evidence refutes A conclusively, and one when there is no evidence against A at all.

As we have defined it, the degree of plausibility is obviously analogous to the upper probability of Chapter 2. Indeed, by adding the phrase *degree of dubiety* to name the degree to which the evidence impugns a proposition, we obtain a complete evidential vocabulary parallel to the subjective

Table 7.1

The Subjective Vocabulary		The Evidential Vocabulary	
Degree of belief	$Bel(A)$	Degree of support	$S(A)$
Degree of doubt	$Dou(A) = Bel(\bar{A})$	Degree of dubiety	$Dub(A) = S(\bar{A})$
Upper probability	$P^*(A) = 1 - Bel(\bar{A})$	Degree of plausibility	$Pl(A) = 1 - S(\bar{A})$

vocabulary of Chapter 2. Since the remainder of this essay is mainly concerned with support functions rather than belief functions in general, it will mainly employ this evidential vocabulary and notation.

A support function $S: 2^{\Theta} \to [0,1]$ is fully described, of course, by the *plausibility function* $Pl: 2^{\Theta} \to [0,1]$ defined by (7.2), and we will often conduct our discussions in terms of Pl.

Degrees of plausibility are of particular interest when one attends to singletons or to other relatively small subsets of a frame of discernment. For the evidence may award such subsets varying plausibilities even though none of them attain positive support.

EXAMPLE 7.2. *An Exhaustive List of Suspects.* Suppose we have a list of suspects for the burglary of the sweetshop, and we know that exactly one of the suspects on the list is the true thief. Then we can take the set of these suspects to be our frame of discernment Θ.* And if we know which of the suspects are left-handed and which are insiders, then we can use this frame to combine the evidence against the two groups. Indeed, let $L \subset \Theta$ be the set of left-handers, and let $I \subset \Theta$ be the set of insiders. And suppose, contrary to the assumption of Example 4.1, that each of the sets $L \cap I$, $L \cap \bar{I}$, $\bar{L} \cap I$ and $\bar{L} \cap \bar{I}$ has many elements. Then if the evidence consists of a weight of $\log 2$ focused on L and another of $\log 2$ focused on I, we obtain a support function $S: 2^{\Theta} \to [0,1]$ with focal elements L, I, $L \cap I$ and Θ, all with basic probability number $\frac{1}{4}$. $S(\{\theta\}) = 0$ for each $\theta \in \Theta$, but $Pl(\{\theta\})$ varies:

$$Pl(\{\theta\}) = \begin{cases} \frac{1}{4} & \text{if } \theta \in \bar{L} \cap \bar{I} \\ \frac{1}{2} & \text{if } \theta \in L \cap \bar{I} \text{ or } \theta \in \bar{L} \cap I \\ 1 & \text{if } \theta \in L \cap I. \blacksquare \end{cases}$$

*Since our frame of discernment is epistemic in nature, we must refrain from blithely proclaiming "Let Θ be the set of all possible thieves" unless we have such a list.

§3. The Vacuous Extension of a Support Function

There are occasions when the impact of a body of evidence on a frame Ω is fully discerned by a coarsening Θ. The support functions on the two frames are then related in a simple way.

What, indeed, does it mean to say that the frame Θ discerns the full impact of the evidence on a refinement Ω? It means that no proposition discerned by Ω attains any greater support than that which it receives by virtue of being implied by propositions discerned by Θ. In other words, the support function S over Ω can be obtained from the support function S_0 over Θ by the relation

$$S(A) = \max_{\substack{B \subset \Theta \\ \omega(B) \subset A}} S_0(B) \tag{7.3}$$

for all $A \subset \Theta$, where $\omega : 2^{\Theta} \to 2^{\Omega}$ is the refining.

Whenever two support functions S and S_0 are related by (7.3), I shall say that S is the *vacuous extension* of S_0 to Ω, and that S is *carried* by the coarsening Θ. As the following theorem confirms, a support function on a given frame can always be vacuously extended to a support function on a refinement.

THEOREM 7.3. *Suppose* $S_0 : 2^{\Theta} \to [0, 1]$ *is a support function, and* $\omega : 2^{\Theta} \to 2^{\Omega}$ *is a refining. Then the function* $S : 2^{\Omega} \to [0, 1]$ *defined by (7.3) is a support function, and* $S_0 = S|2^{\Theta}$.

There are also several other ways to characterize the vacuous extension of a support function:

THEOREM 7.4. *Suppose* $\omega : 2^{\Theta} \to 2^{\Omega}$ *is a refining, and let* $\underline{\theta}$ *and* $\overline{\theta}$ *denote its inner and outer reductions. Suppose* $S_0 : 2^{\Theta} \to [0, 1]$ *and* $S : 2^{\Omega} \to [0, 1]$ *are support functions, and let* m_0, m, Pl_0, Pl, Q_0 *and* Q *denote their corresponding basic probability assignments, plausibility functions, and commonality functions. Then the following conditions are all equivalent:*

(1) S *is the vacuous extension of* S_0 *to* Ω.

(2) $Pl(A) = \min_{\substack{B \subset \Theta \\ A \subset \omega(B)}} Pl_0(B)$ *for all* $A \subset \Omega$.

(3) $m_0 = m \circ \omega$.

(4) $S = S_0 \circ \underline{\theta}$.

(5) $Pl = Pl_0 \circ \overline{\theta}$.

(6) $Q = Q_0 \circ \overline{\theta}$.

THEOREM 7.5. *Suppose* Ω *is a refinement of* Θ, $S_0 : 2^\Theta \to [0,1]$ *is a support function,* $S : 2^\Omega \to [0,1]$ *is the vacuous extension of* S_0, *and* $S' : 2^\Omega \to [0,1]$ *is another support function such that* $S_0 = S'|2^\Theta$. *Denote the commonality functions for* S *and* S' *by* Q *and* Q', *respectively. Then*

$$Q(A) \geq Q'(A)$$

for all $A \subset \Omega$.

We may interpret Theorem 7.5 by saying that of all the support functions over Ω which yield S_0 when restricted to 2^Θ, the vacuous extension imposes the least constraint on the movement of one's probability masses.

Vacuous extensions will often arise in the evaluation of evidence. Given a fixed body of evidence $\mathcal{E}$ and a frame of discernment Θ, one will usually be able to find a refinement Ω such that $\mathcal{E}$ has no impact on Ω save what is already discerned by Θ. One should not suppose, though, that this could happen for *all* refinements of Θ; it may sometimes be impossible to exhaust the richness of a given body of evidence with any single fixed frame of discernment.

§4. Evidential Independence

As we learned earlier, two compatible frames are independent if no proposition discerned by one of them non-trivially implies a proposition discerned by the other. Another way of putting this is to say that learning whether a proposition discerned by one frame is true or false will not in itself result in learning whether a proposition discerned by the other is true or false. Does the independence of two frames allow one to make any analogous statements about evidence that falls short of establishing truth or falsehood?

Indeed it does. We notice first of all that if the impact of a body of evidence on a frame of discernment Ω is fully discerned by a coarsening Θ_1, then that evidence should not induce any positive degrees of support for propositions discerned by an independent frame Θ_2.

THEOREM 7.6. *Suppose* Θ_1 *and* Θ_2 *are independent coarsenings of a frame of discernment* Ω. *And suppose the support function* S *over* Ω *is carried by* Θ_1. *Then* $S|2^{\Theta_2}$ *is vacuous.*

More generally, if a support function is carried by one coarsening, then its values for that coarsening will not change when it is combined with a support function carried by an independent coarsening:

THEOREM 7.7. *Suppose* Θ_1 *and* Θ_2 *are independent coarsenings of a frame* Ω. *And suppose the support functions* S_1 *and* S_2 *over* Ω *are carried by* Θ_1 *and* Θ_2, *respectively. Then*
$(S_1 \oplus S_2)|2^{\Theta_1} = S_1|2^{\Theta_1}$.

When the degrees of suppórt for a frame of discernment are obtained by combining support functions carried by two independent coarsenings, those coarsenings seem to be more than merely logically independent; they are independent with respect to the evidence. Formally, we may say that two independent frames of discernment Θ_1 and Θ_2 are *evidentially independent* with respect to a support function S over their minimal refinement $\Theta_1 \otimes \Theta_2$ if S can be obtained by combining a support function over $\Theta_1 \otimes \Theta_2$ that is carried by Θ_1 with one that is carried by Θ_2. And more generally, we may say that two independent frames Θ_1 and Θ_2 are *evidentially independent* with respect to a support function S over any given common refinement of Θ_1 and Θ_2 if they are evidentially independent with respect to $S|2^{\Theta_1 \otimes \Theta_2}$.

The following theorem provides two alternatives to this intuitive but clumsy definition of evidential independence:

THEOREM 7.8. *Suppose* $S: 2^{\Omega} \to [0,1]$ *is a support function and* Θ_1 *and* Θ_2 *are independent coarsenings of* Ω. *Denote* S's *plausibility function by* Pl. *Then the following assertions are all equivalent:*

(1) Θ_1 *and* Θ_2 *are evidentially independent with respect to* S.

(2) $S|2^{\Theta_1 \otimes \Theta_2} = S_1 \oplus S_2$, *where* S_i *is the vacuous extension to* $\Theta_1 \otimes \Theta_2$ *of* $S|2^{\Theta_i}$, *for* $i = 1, 2$.

(3) *Both*

$$S(A \cap B) = S(A)S(B)$$

and

$$Pl(A \cap B) = Pl(A)Pl(B)$$

whenever A *is discerned by* Θ_1 *and* B *is discerned by* Θ_2.

All the ideas of this section can, of course, be generalized to the case of n independent frames of discernment.

§5. Cognitive Independence

The notion of evidential independence is retrospective: two frames of discernment are evidentially independent with respect to a support function if that support function could be obtained by combining evidence that bears on only one of them with evidence that bears on only the other. But there is a second notion of independence that is prospective: two frames of discernment may be called *cognitively independent* with respect to the evidence if new evidence that bears on only one of them will not change the degrees of support for propositions discerned by the other.

More precisely, two independent coarsenings Θ_1 and Θ_2 of a frame of discernment Ω are said to be *cognitively independent* with respect to a support function S over Ω if

$$(S \oplus S_1)|2^{\Theta_2} = S|2^{\Theta_2}$$

for every support function S_1 over Ω that is carried by Θ_1 and is combinable with S.

Cognitive independence is in fact a weaker condition than evidential independence: if two frames are evidentially independent with respect to a support function, then they will be cognitively independent with respect to it. This fact may be deduced from Theorem 7.7 together with the definitions, but it is more immediately obvious from the following theorem:

THEOREM 7.9. *Suppose Θ_1 and Θ_2 are independent coarsenings of a frame Ω, and S is a belief function over Ω. Denote S's plausibility function by Pl. Then Θ_1 and Θ_2 are cognitively independent with respect to S if and only if*

$$Pl(A \cap B) = Pl(A)\,Pl(B)$$

whenever A is discerned by Θ_1 and B is discerned by Θ_2.

So while evidential independence requires both

$$S(A \cap B) = S(A)\,S(B) \tag{7.4}$$

and

$$Pl(A \cap B) = Pl(A)\,Pl(B) \tag{7.5}$$

whenever A is discerned by Θ_1 and B by Θ_2, cognitive independence requires only (7.5).

Cognitive independence may hold even though evidential independence fails, for (7.5) does not imply (7.4). In fact, neither (7.4) nor (7.5) implies the other. This may be established by exhibiting two simple examples.

EXAMPLE 7.3. *Cognitive Independence.* Set

$$\Omega = \{a, b, c, d\}, \ \Theta_1 = \{\alpha, \beta\} \text{ and } \Theta_2 = \{\gamma, \delta\} .$$

Define the refining $\omega_1 : 2^{\Theta_1} \to 2^{\Omega}$ by

$$\omega_1(\{\alpha\}) = \{a, c\} \equiv A, \ \omega_1(\{\beta\}) = \{b, d\} \equiv B ,$$

and define the refining $\omega_2 : 2^{\Theta_2} \to 2^{\Omega}$ by

$$\omega_2(\{\gamma\}) = \{a, b\} \equiv \Gamma, \ \omega_2(\{\delta\}) = \{c, d\} \equiv \Delta .$$

Then Θ_1 and Θ_2 are independent coarsenings of Ω.

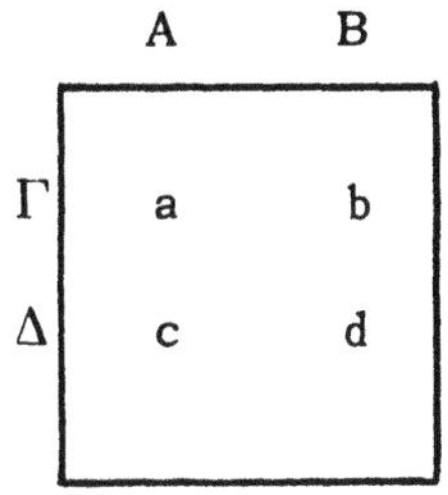

Figure 7.1

(1) Define a support function S over Ω by setting the basic probability numbers for $\{a, b, c\}$, A, Γ and $\{a\}$ all equal to $\frac{1}{4}$. Then

$$\frac{1}{4} = S(\{a\}) = S(A \cap \Gamma) = S(A)\,S(\Gamma) = \frac{1}{2} \cdot \frac{1}{2} = \frac{1}{4} ,$$

$$0 = S(\{b\}) = S(B \cap \Gamma) = S(B)\,S(\Gamma) = 0 \cdot \frac{1}{2} = 0 ,$$

$$0 = S(\{c\}) = S(A \cap \Delta) = S(A)\,S(\Delta) = \frac{1}{2} \cdot 0 = 0 ,$$

$$0 = S(\{d\}) = S(B \cap \Delta) = S(B)\,S(\Delta) = 0 \cdot 0 = 0 ,$$

but

$$0 = \mathrm{Pl}(\{d\}) = \mathrm{Pl}(B \cap \Delta) \neq \mathrm{Pl}(B)\,\mathrm{Pl}(\Delta) = \frac{1}{2} \cdot \frac{1}{2} = \frac{1}{4} .$$

(2) Define a support function S over Ω by setting the basic probability numbers for Ω and $\{a, b, c\}$ equal to $\frac{1}{4}$ and the one for $\{a\}$ equal to $\frac{1}{2}$. Then

$$1 = \text{Pl}(\{a\}) = \text{Pl}(A \cap \Gamma) = \text{Pl}(A)\,\text{Pl}(\Gamma) = 1 \cdot 1 = 1\,,$$

$$\frac{1}{2} = \text{Pl}(\{b\}) = \text{Pl}(B \cap \Gamma) = \text{Pl}(B)\,\text{Pl}(\Gamma) = \frac{1}{2} \cdot 1 = \frac{1}{2}\,,$$

$$\frac{1}{2} = \text{Pl}(\{c\}) = \text{Pl}(A \cap \Delta) = \text{Pl}(A)\,\text{Pl}(\Delta) = 1 \cdot \frac{1}{2} = \frac{1}{2}\,,$$

$$\frac{1}{4} = \text{Pl}(\{d\}) = \text{Pl}(B \cap \Delta) = \text{Pl}(B)\,\text{Pl}(\Delta) = \frac{1}{2} \cdot \frac{1}{2} = \frac{1}{4}\,,$$

but

$$\frac{1}{2} = S(\{a\}) = S(A \cap \Gamma) \neq S(A)\,S(\Gamma) = \frac{1}{2} \cdot \frac{1}{2} = \frac{1}{4}\,. \ \blacksquare$$

§6. Mathematical Appendix

LEMMA 7.1. *Suppose* $\text{Bel}_1, \cdots, \text{Bel}_n$ *are belief functions over a common frame* Θ, *and denote their basic probability assignments by* $m_1, \cdots, m_n$. *Suppose* $\text{Bel}_1 \oplus \cdots \oplus \text{Bel}_n$ *exists, and denote its basic probability assignment by* m. *Then*

$$m(A) = \frac{\displaystyle\sum_{\substack{A_1, \cdots, A_n \subset \Theta \\ A_1 \cap \cdots \cap A_n = A}} m_1(A_1) \cdots m_n(A_n)}{\displaystyle\sum_{\substack{A_1, \cdots, A_n \subset \Theta \\ A_1 \cap \cdots \cap A_n \neq \emptyset}} m_1(A_1) \cdots m_n(A_n)}$$

for all non-empty subsets of Θ.

Proof of Lemma 7.1. Let us prove the lemma by induction on n, using Theorem 3.1.

The case $n = 2$ is immediate by Theorem 3.1.

Let us assume the lemma is true for $n = k$, as well as for $n = 2$, and then prove it for $n = k+1$. Let m_0 denote the basic probability assignment for $Bel_1 \oplus \cdots \oplus Bel_k$. Then the basic probability assignment for $Bel_1 \oplus \cdots \oplus Bel_{k+1}$ is given by

$$m(A) = \frac{\displaystyle\sum_{\substack{A_0, A_{k+1} \subset \Theta \\ A_0 \cap A_{k+1} = A}} m_0(A_0)\, m_{k+1}(A_{k+1})}{\displaystyle\sum_{\substack{A_0, A_k \subset \Theta \\ A_0 \cap A_{k+1} \neq \emptyset}} m_0(A_0)\, m_{k+1}(A_{k+1})}$$

$$= \frac{\displaystyle\sum_{\substack{A_0, A_{k+1} \subset \Theta \\ A_0 \cap A_{k+1} = A}} \left(\sum_{\substack{A_1, \cdots, A_k \subset \Theta \\ A_1 \cap \cdots \cap A_k = A_0}} m_1(A_1) \cdots m_k(A_k) \right) m_{k+1}(A_{k+1})}{\displaystyle\sum_{\substack{A_0, A_{k+1} \subset \Theta \\ A_0 \cap A_{k+1} \neq \emptyset}} \left(\sum_{\substack{A_1, \cdots, A_k \subset \Theta \\ A_1 \cap \cdots \cap A_k = A_0}} m_1(A_1) \cdots m_k(A_k) \right) m_{k+1}(A_{k+1})}$$

$$= \frac{\displaystyle\sum_{\substack{A_1, \cdots, A_{k+1} \subset \Theta \\ A_1 \cap \cdots \cap A_{k+1} = A}} m_1(A_1) \cdots m_{k+1}(A_{k+1})}{\displaystyle\sum_{\substack{A_1, \cdots, A_{k+1} \subset \Theta \\ A_1 \cap \cdots \cap A_{k+1} \neq \emptyset}} m_1(A_1) \cdots m_{k+1}(A_{k+1})}$$

for all non-empty $A \subset \Theta$. ■

LEMMA 7.2. *Suppose* $S_1, \cdots, S_n$ *are simple support functions over a frame* Θ. *Let* A_i *denote the focus of* S_i, *and set* $S_i(A_i) = s_i$. *Suppose* $S = S_1 \oplus \cdots \oplus S_n$ *exists, and let* m *denote its basic probability assignment. Then*

$$m(A) = \frac{\displaystyle\sum_{\substack{I \subset \{1,\cdots,n\} \\ \bigcap_{i \in I} A_i = A}} \Big(\prod_{i \in I} s_i\Big)\Big(\prod_{i \in \bar{I}} (1-s_i)\Big)}{\displaystyle\sum_{\substack{I \subset \{1,\cdots,n\} \\ \bigcap_{i \in I} A_i \neq \emptyset}} \Big(\prod_{i \in I} s_i\Big)\Big(\prod_{i \in \bar{I}} (1-s_i)\Big)} \tag{7.6}$$

for all non-empty $A \subset \Theta$. *(Here* $\bar{I} = \{1,\cdots,n\} - I$, $\prod_{i \in \emptyset} s_i = \prod_{i \in \emptyset} (1-s_i) = 1$, *and* $\bigcap_{i \in \emptyset} A_i = \Theta$.)

Proof of Lemma 7.2. S_i has the focal elements A_i and Θ, with $m_i(A_i) = s_i$ and $m_i(\Theta) = 1 - s_i$. Hence the formula of Lemma 7.1 reduces to (7.6). ■

LEMMA 7.3. *The commonality number of the core of a belief function is equal to its basic probability number.*

Proof of Lemma 7.3. Suppose, indeed, that $\text{Bel}: 2^\Theta \to [0, 1]$ is a belief function, with core $\mathcal{C}$, basic probability assignment m and commonality function Q. We know that $m(A) = 0$ for all $A \subset \Theta$ that are not contained in $\mathcal{C}$. Hence

$$Q(\mathcal{C}) = \sum_{\substack{A \subset \Theta \\ \mathcal{C} \subset A}} m(A) = \sum_{\substack{A \subset \Theta \\ \mathcal{C} \subset A \subset \mathcal{C}}} m(A) = m(\mathcal{C}) \,. \; ■$$

LEMMA 7.4. *Suppose* Θ *is a frame of discernment, and* $B_1, \cdots, B_n$ *are subsets of* Θ. *Then there exists a refinement* Ω *of* Θ, *with outer reduction* $\bar{\theta}: 2^\Omega \to 2^\Theta$, *and subsets* $A_1, \cdots, A_n$ *of* Ω *such that* $\bar{\theta}(A_i) = B_i$ *for all* i, *and* $A_i \cap A_j = \emptyset$ *when* $i \neq j$.

Proof of Lemma 7.4. By condition (5) of the definition of a compatible family, there exists, for any given element $\theta \in \Theta$, a refining that maps θ

to a set of n elements. It follows that there exists a refining $\omega : 2^{\Theta} \to 2^{\Omega}$ such that $|\omega(\{\theta\})| \geq n$ for all $\theta \in \Theta$. Given such a refining ω, consider each $\omega(\{\theta\})$, and enumerate n of its elements, denoting them $\omega_1^{\theta}, \cdots, \omega_n^{\theta}$. Then for each i, $i = 1, \cdots, n$, set

$$A_i = \{\omega_i^{\theta} | \theta \in B_i\} \subset \Omega .$$

It is evident that the A_i satisfy the requirements of the lemma. ■

Proof of Theorem 7.1. The equivalence of (2) and (3) follows immediately from Lemma 7.3. Hence it suffices to show that (1) and (3) are equivalent.

(i) Assume first that Bel is a support function – that Bel = $S|2^{\Theta}$, where S is a separable support function over some refinement Ω of Θ. Then we must show that $m_0(\mathcal{C}) > 0$, where m_0 is Bel's basic probability assignment. Let $\bar{\theta} : 2^{\Omega} \to 2^{\Theta}$ denote the outer reduction, let m denote the basic probability assignment for S, and let $\mathcal{C}_S$ denote the core of S. Then by Theorem 5.3, $m(\mathcal{C}_S) > 0$. And by Theorem 6.9, $\mathcal{C} = \bar{\theta}(\mathcal{C}_S)$, and

$$m_0(\mathcal{C}) = \sum_{\substack{B \subset \Omega \\ \mathcal{C} = \bar{\theta}(B)}} m(B) \geq m(\mathcal{C}_S) > 0 ,$$

as required.

(ii) Now let us assume that $m_0(\mathcal{C}) > 0$, still letting m_0 denote Bel's basic probability assignment. Then we must show that Bel is a support function – we must construct a separable support function S over a refinement Ω of Θ such that Bel = $S|2^{\Theta}$. This can be done as follows. Let $B_1, \cdots, B_n$ denote the other focal elements of Bel, so that

$$m_0(\mathcal{C}) + \sum_{i=1}^{n} m_0(B_i) = 1 .$$

Now suppose $\omega : 2^{\Theta} \to 2^{\Omega}$ is a refining, and the subsets $A_1, \cdots, A_n$ of Ω satisfy the requirements of Lemma 7.4. For each i, $i = 1, \cdots, n$, let $S_i : 2^{\Omega} \to [0, 1]$ be the simple support function focused on A_i, with

$$S_i(A_i) = \frac{m_0(B_i)}{m_0(B_i) + m_0(\mathcal{C})} < 1 .$$

Now set $A_{n+1} = \omega(\mathcal{C})$. If $\mathcal{C} = \Theta$, then $A_{n+1} = \Omega$, but if $\mathcal{C}$ is a proper subset of Θ, then A_{n+1} will be a proper subset of Ω. In the latter case, let $S_{n+1} : 2^{\Omega} \to [0,1]$ denote the simple support function focused on A_{n+1} with $S_{n+1}(A_{n+1}) = 1$.

Now define a support function S over Ω by setting $S = S_1 \oplus \cdots \oplus S_n$ if $\mathcal{C} = \Theta$ and $S = S_1 \oplus \cdots \oplus S_{n+1}$ if $\mathcal{C} \neq \Theta$. And let m denote the basic probability assignment for S. If we set $s_i = S_i(A_i)$ for each i and use the fact that $A_1, \cdots, A_n$ are disjoint, then we can calculate the values of m by (7.6). Indeed,

$$m(A_i) = \frac{s_i \prod_{j \neq i, n+1} (1 - s_j)}{\prod_{j=1}^{n} (1 - s_j) + \sum_{j=1}^{n} s_j \prod_{k \neq j, n+1} (1 - s_k)}$$

$$= \frac{\dfrac{s_i}{1 - s_i}}{1 + \sum_{j=1}^{n} \dfrac{s_j}{1 - s_j}}$$

$$= m_0(B_i)$$

for $i = 1, \cdots, n$,

$$m(A_{n+1}) = \frac{\prod_{i=1}^{n} (1 - s_i)}{\prod_{i=1}^{n} (1 - s_i) + \sum_{i=1}^{n} s_i \prod_{j \neq i, n+1} (1 - s_j)}$$

$$= \frac{1}{1 + \sum_{i=1}^{n} \dfrac{s_i}{1 - s_i}}$$

$$= m_0(\mathcal{C}) ,$$

and $m(A) = 0$ if A is not equal to any of the A_i. (This same result is obtained whether or not $\mathcal{C} = \Theta$.)

Now according to Theorem 6.9, the basic probability number that $S|2^{\Theta}$ assigns to $A \subset \Theta$ is given by

$$\sum_{\substack{B \subset \Omega \\ A = \bar{\theta}(B)}} m(B) = \sum_{\substack{i \\ A = \bar{\theta}(A_i)}} m(A_i) .$$

Since $\bar{\theta}(A_i) = B_i$ for $i = 1,\cdots,n$, and $\bar{\theta}(A_{n+1}) = \mathcal{C}$, and since the B_i are all distinct from each other and from $\mathcal{C}$, this reduces to $m_0(B_i)$ if $A = B_i$, to $m_0(\mathcal{C})$ if $A = \mathcal{C}$, and to zero otherwise. So m_0 is the basic probability assignment for $S|2^{\Theta}$. In other words, $Bel = S|2^{\Theta}$. ∎

Proof of Theorem 7.2. We may suppose that $S = S_1 \oplus \cdots \oplus S_n$, where the S_i are simple support functions, each of which accords positive support s_i to its focus A_i. Then we see by Lemma 7.2 that A and B are focal elements of S only if there exist subsets I and J of $\{1,\cdots,n\}$ such that

$$\bigcap_{i \in I} A_i = A, \quad \bigcap_{i \in J} B_i = B ,$$

and $s_i < 1$ if $i \in \bar{I}$ or $i \in \bar{J}$. If this is so,

$$\bigcap_{i \in (I \cup J)} A_i = A \cap B ,$$

and $s_i < 1$ if $i \in \overline{I \cup J}$. If $A \cap B \neq \emptyset$, then it follows, again by Lemma 7.2, that $A \cap B$ is a focal element of S. ∎

LEMMA 7.5. *Suppose* Bel *is a belief function over* Θ, *and suppose* $\omega : 2^{\Theta} \to 2^{\Omega}$ *is a refining. Let* P^* *denote the upper probability function for* Bel, *and let* $\underline{\theta}$ *and* $\bar{\theta}$ *denote the inner and outer reductions for* ω. *Then* $Bel \circ \underline{\theta}$ *is a belief function over* Ω, *and* $P^* \circ \bar{\theta}$ *is its upper probability function. These two functions are given by*

$$(\mathrm{Bel}\circ\underline{\theta})(A) = \max_{\substack{B\subset\Theta\\ \omega(B)\subset A}} \mathrm{Bel}(B)$$

and

$$(P^*\circ\overline{\theta})(A) = \min_{\substack{B\subset\Theta\\ A\subset\omega(B)}} P^*(B)$$

for all $A\subset\Omega$.

Proof of Lemma 7.5. First let us prove that $\mathrm{Bel}\circ\underline{\theta}$ is a belief function:

(1) $(\mathrm{Bel}\circ\underline{\theta})(\emptyset) = \mathrm{Bel}(\underline{\theta}(\emptyset)) = \mathrm{Bel}(\emptyset) = 0$.

(2) $(\mathrm{Bel}\circ\underline{\theta})(\Omega) = \mathrm{Bel}(\underline{\theta}(\Omega)) = \mathrm{Bel}(\Theta) = 1$.

(3) According to (7) of Theorem 6.3, $\underline{\theta}(A)\subset\underline{\theta}(B)$ whenever $A\subset B\subset\Omega$. It follows that $\underline{\theta}(A_1)\cup\cdots\cup\underline{\theta}(A_n)\subset\underline{\theta}(A_1\cup\cdots\cup A_n)$ whenever $A_1,\cdots,A_n$ are subsets of Ω. By the definition of a belief function, $\mathrm{Bel}(A)\le\mathrm{Bel}(B)$ whenever $A\subset B$, so we may conclude that $\mathrm{Bel}(\underline{\theta}(A_1)\cup\cdots\cup\underline{\theta}(A_n))\le\mathrm{Bel}(\underline{\theta}(A_1\cup\cdots\cup A_n))$. So by (8) of Theorem 6.3,

$$\begin{aligned}(\mathrm{Bel}\circ\underline{\theta})(A_1\cup\cdots\cup A_n) &\ge \mathrm{Bel}(\underline{\theta}(A_1)\cup\cdots\cup\underline{\theta}(A_n))\\ &\ge \sum_{\substack{I\subset\{1,\cdots,n\}\\ I\ne\emptyset}} (-1)^{|I|+1}\,\mathrm{Bel}\Big(\bigcap_{i\in I}\underline{\theta}(A_i)\Big)\\ &= \sum_{\substack{I\subset\{1,\cdots,n\}\\ I\ne\emptyset}} (-1)^{|I|+1}\,(\mathrm{Bel}\circ\underline{\theta})\Big(\bigcap_{i\in I}A_i\Big)\end{aligned}$$

whenever $A_1,\cdots,A_n$ are subsets of Ω.

By (10) of Theorem 6.3,

$$\begin{aligned}(P^*\circ\overline{\theta})(A) &= P^*(\overline{\theta}(A)) = 1-\mathrm{Bel}(\overline{\overline{\theta}(A)})\\ &= 1-\mathrm{Bel}(\underline{\theta}(\bar{A}))\\ &= 1-(\mathrm{Bel}\circ\underline{\theta})(\bar{A}),\end{aligned}$$

whence $P^*\circ\overline{\theta}$ is indeed the upper probability function for $\mathrm{Bel}\circ\underline{\theta}$.

By Theorem 6.2, $\omega(B) \subset A$ if and only if $B \subset \underline{\theta}(A)$. Since $\mathrm{Bel}(B) \leq \mathrm{Bel}(\underline{\theta}(A))$ whenever $B \subset \underline{\theta}(A)$, it follows that

$$\max_{\substack{B \subset \Theta \\ \omega(B) \subset A}} \mathrm{Bel}(B) = \max_{\substack{B \subset \Theta \\ B \subset \underline{\theta}(A)}} \mathrm{Bel}(B) = \mathrm{Bel}(\underline{\theta}(A))$$

for all $A \subset \Omega$. Similarly, $A \subset \omega(B)$ if and only if $\overline{\theta}(A) \subset B$. Since $P^*(\overline{\theta}(A)) \leq P^*(B)$ whenever $\overline{\theta}(A) \subset B$, it follows that

$$\min_{\substack{B \subset \Theta \\ A \subset \omega(B)}} P^*(B) = \min_{\substack{B \subset \Theta \\ \overline{\theta}(A) \subset B}} P^*(B) = P^*(\overline{\theta}(A))$$

for all $A \subset \Omega$. ■

LEMMA 7.6. *Suppose* $\omega : 2^{\Theta} \to 2^{\Omega}$ *is a refining, and let* $\underline{\theta}$ *and* $\overline{\theta}$ *denote its inner and outer reductions. Suppose* $\mathrm{Bel}_0 : 2^{\Theta} \to [0,1]$ and $\mathrm{Bel} : 2^{\Omega} \to [0,1]$ *are belief functions, and let* m_0, m, P_0^*, P^*, Q_0, *and* Q *denote their corresponding basic probability assignments, upper probability functions, and commonality functions. Then the following conditions are all equivalent*:

(1) $\mathrm{Bel} = \mathrm{Bel}_0 \circ \underline{\theta}$.

(2) $m_0 = m \circ \omega$.

(3) $P^* = P_0^* \circ \overline{\theta}$.

(4) $Q = Q_0 \circ \overline{\theta}$.

Proof of Lemma 7.6. (1) implies (2). Suppose $\mathrm{Bel} = \mathrm{Bel}_0 \circ \underline{\theta}$. Define $m' : 2^{\Omega} \to [0,1]$ by

$$m'(C) = \begin{cases} m_0(B) & \text{if } C = \omega(B) \\ 0 & \text{if } C \neq \omega(B) \text{ for any } B \subset \Theta. \end{cases}$$

Since ω is one-to-one, m' is well defined, and

$$\sum_{C \subset \Omega} m'(C) = \sum_{B \subset \Theta} m_0(B) = 1 ,$$

whence m' is a basic probability assignment. Notice that $m' \circ \omega = m_0$. And

$$\mathrm{Bel}(A) = \mathrm{Bel}_0(\underline{\theta}(A)) = \sum_{B \subset \underline{\theta}(A)} m_0(B)$$

$$= \sum_{\substack{B \subset \Theta \\ \omega(B) \subset A}} m_0(B)$$

$$= \sum_{C \subset A} m'(C) \,.$$

So $m' = m$ by Theorem 2.2, and hence $m_0 = m' \circ \omega = m \circ \omega$.

(2) implies (1). Suppose $m_0 = m \circ \omega$. Then

$$1 = \sum_{A \subset \Theta} m_0(A) = \sum_{A \subset \Theta} m(\omega(A)) \,,$$

whence $m(B) = 0$ if $B \neq \omega(A)$ for any $A \subset \Theta$. Thus

$$\mathrm{Bel}(A) = \sum_{B \subset A} m(B) = \sum \{m(B) | B \subset A;\ B = \omega(C) \text{ for some } C \subset \Theta\}$$

$$= \sum \{m(\omega(C)) | C \subset \Theta;\ \omega(C) \subset A\}$$

$$= \sum \{m_0(C) | C \subset \Theta;\ C \subset \underline{\theta}(A)\}$$

$$= \mathrm{Bel}_0(\underline{\theta}(A))$$

for all $A \subset \Omega$; $\mathrm{Bel} = \mathrm{Bel}_0 \circ \underline{\theta}$.

(3) is equivalent to (1). This follows immediately from Lemma 7.5.

(2) implies (4). Suppose $m_0 = m \circ \omega$. Then we may conclude, as in the second paragraph above, that $m(B) = 0$ if $B \neq \omega(A)$ for any $A \subset \Theta$ And

$$Q(A) = \sum_{\substack{B \\ A \subset B}} m(B) = \sum \{m(B) | A \subset B;\ B = \omega(C) \text{ for some } C \subset \Theta\}$$

$$= \sum \{m(\omega(C)) | C \subset \Theta;\ A \subset \omega(C)\}$$

$$= \sum \{m_0(C) | C \subset \Theta;\ \bar{\theta}(A) \subset C\}$$

$$= Q_0(\bar{\theta}(A))$$

for all $A \subset \Omega$; $Q = Q_0 \circ \bar{\theta}$.

(4) implies (1). By the preceding paragraph, $Q_0 \circ \bar{\theta}$ is the commonality function for $\mathrm{Bel}_0 \circ \underline{\theta}$. So if $Q = Q_0 \circ \bar{\theta}$, then $\mathrm{Bel} = \mathrm{Bel}_0 \circ \underline{\theta}$. ∎

Proof of Theorem 7.3. According to Lemma 7.5, S is a belief function and $S = S_0 \circ \underline{\theta}$, where $\underline{\theta}$ is the inner reduction for ω. By definition, $S|2^\Theta = S \circ \omega$. So $S|2^\Theta = S_0 \circ \underline{\theta} \circ \omega$. But $\underline{\theta} \circ \omega$ is the identity, by (2) of Theorem 6.3. Hence $S|2^\Theta = S_0$.

We must still prove that S is a support function – i.e., that its core has a positive commonality number. But this can be proven by using the same property for S_0. Let m, m_0, $\mathcal{C}$, and $\mathcal{C}_0$ denote the basic probability assignments and cores for S and S_0. Then $m_0 = m \circ \omega$ by Lemma 7.6. So

$$1 = \sum_{A \subset \Theta} m_0(A) = \sum_{A \subset \Theta} m(\omega(A)) ,$$

and it follows that $m(B) = 0$ if $B \neq \omega(A)$ for any $A \subset \Theta$. Hence

$$\begin{aligned} \mathcal{C} &= \cup \{B \subset \Omega | m(B) > 0\} \\ &= \cup \{\omega(A) | A \subset \Theta;\ m_0(A) > 0\} \\ &= \omega(\cup \{A \subset \Theta | m_0(A) > 0\}) \\ &= \omega(\mathcal{C}_0) . \end{aligned}$$

And $m(\mathcal{C}) = (m \circ \omega)(\mathcal{C}_0) = m_0(\mathcal{C}_0) > 0$. ∎

Proof of Theorem 7.4. This theorem is an immediate consequence of Lemmas 7.5 and 7.6. ■

Proof of Theorem 7.5. Let Q_0 denote the commonality function for S_0. Then by Theorem 6.9, $Q_0(A) \geq Q'(\omega(A))$ for all $A \subset \Theta$. By Theorem 7.4, $Q = Q_0 \circ \bar{\theta}$. Hence

$$Q(A) = Q_0(\bar{\theta}(A)) \geq Q'(\omega(\bar{\theta}(A)))$$
$$\geq Q'(A)$$

for all $A \subset \Omega$; the last inequality follows from (3) of Theorem 6.3, together with the monotonicity of Q'. ■

Proof of Theorem 7.6. Denote the refinings by $\omega_1 : 2^{\Theta_1} \to 2^{\Omega}$ and $\omega_2 : 2^{\Theta_2} \to 2^{\Omega}$, and denote the corresponding inner reductions by $\underline{\theta}_1$ and $\underline{\theta}_2$. Then

$$S = (S|2^{\Theta_1}) \circ \underline{\theta}_1$$
$$= S \circ \omega_1 \circ \underline{\theta}_1$$

by hypothesis, so

$$S|2^{\Theta_2} = S \circ \omega_2 = S \circ \omega_1 \circ \underline{\theta}_1 \circ \omega_2 .$$

But by (5) of Theorem 6.10,

$$(\underline{\theta}_1 \circ \omega_2)(A) = \emptyset$$

whenever $A \subset \Theta_2$, $A \neq \Theta_2$. So

$$(S|2^{\Theta_2})(A) = (S \circ \omega_1 \circ \underline{\theta}_1 \circ \omega_2)(A) = S(\omega_1(\emptyset)) = S(\emptyset) = 0$$

whenever $A \subset \Theta_2$, $A \neq \Theta_2$; $S|2^{\Theta_2}$ is vacuous. ■

LEMMA 7.7. *Suppose* $\omega_1 : 2^{\Theta_1} \to 2^{\Omega}$ *and* $\omega_2 : 2^{\Theta_2} \to 2^{\Omega}$ *are refinings, and* Θ_1 *and* Θ_2 *are independent. Suppose* A_1 *and* B_1 *are non-empty subsets of* Θ_1, *and suppose* A_2 *and* B_2 *are non-empty subsets of* Θ_2. *Then*

(1) $\omega_1(A_1) \cap \omega_2(A_2) \subset \omega_1(B_1) \cap \omega_2(B_2)$ *if and only if* $A_1 \subset B_1$ *and* $A_2 \subset B_2$;

(2) $(\omega_1(A_1) \cap \omega_2(A_2)) \cap (\omega_1(B_1) \cap \omega_2(B_2)) \neq \emptyset$ *if and only if* $A_1 \cap B_1 \neq \emptyset$ *and* $A_2 \cap B_2 \neq \emptyset$;

(3) $\omega_1(A_1) \cap \omega_2(A_2) = \omega_1(B_1) \cap \omega_2(B_2)$ *if and only if* $A_1 = B_1$ *and* $A_2 = B_2$.

Proof of Lemma 7.7.

(1) consider the subsets of Ω of the form

$$\omega_1(\{\theta_1\}) \cap \omega_2(\{\theta_2\}) ,$$

where $\theta_1 \in \Theta_1$ and $\theta_2 \in \Theta_2$. By Theorem 6.10, these subsets are all non-empty, and by Lemma 6.2, they are disjoint. Now whenever $A_1 \subset \Theta_1$ and $A_2 \subset \Theta_2$,

$$\omega_1(A_1) \cap \omega_2(A_2) = \left(\bigcup_{\theta_1 \in A_1} \omega_1(\{\theta_1\})\right) \cap \left(\bigcup_{\theta_2 \in A_2} \omega_2(\{\theta_2\})\right)$$

$$= \bigcup_{\theta_1 \in A_1, \theta_2 \in A_2} (\omega_1(\{\theta_1\}) \cap \omega_2(\{\theta_2\})) .$$

Assertion (1) follows.

(2) $(\omega_1(A_1) \cap \omega_2(A_2)) \cap (\omega_1(B_1) \cap \omega_2(B_2)) = \omega_1(A_1 \cap B_1) \cap \omega_2(A_2 \cap B_2)$,

so (2) follows by (3) of Theorem 6.10.

(3) follows immediately from (1). ∎

LEMMA 7.8. *Suppose* Θ_1 *and* Θ_2 *are independent coarsenings of* Ω, *and denote their corresponding refinings and inner reductions by* ω_1, ω_2, $\underline{\theta}_1$ *and* $\underline{\theta}_2$. *Suppose* $\mathrm{Bel}_1 : 2^{\Theta_1} \to [0,1]$ *and* $\mathrm{Bel}_2 : 2^{\Theta_2} \to [0,1]$ *are belief functions, and denote their basic probability assignments by* m_1 *and* m_2. *Then* $(\mathrm{Bel}_1 \circ \underline{\theta}_1) \oplus (\mathrm{Bel}_2 \circ \underline{\theta}_2)$ *exists, and its basic probability assignment* m *satisfies*

$$m(\omega_1(A) \cap \omega_2(B)) = m_1(A) m_2(B)$$

whenever $A \subset \Theta_1$ *and* $B \subset \Theta_2$, *and* $m(C) = 0$ *if there do not exist* $A \subset \Theta_1$ *and* $B \subset \Theta_2$ *such that* $C = \omega_1(A) \cap \omega_2(B)$.

Proof of Lemma 7.8. Let $A_1, \cdots, A_k$ and $B_1, \cdots, B_\ell$ denote the focal elements of Bel_1 and Bel_2, respectively. Let m'_1 and m'_2 denote the basic probability assignments for $Bel_1 \circ \underline{\theta}_1$ and $Bel_2 \circ \underline{\theta}_2$, respectively. According to Lemma 7.6, $m_1 = m'_1 \circ \omega_1$ and $m_2 = m'_2 \circ \omega_2$. This means in particular that $\omega_1(A_1), \cdots, \omega_1(A_k)$ are the focal elements of $Bel_1 \circ \underline{\theta}_1$ and $\omega_2(B_1), \cdots, \omega_2(B_\ell)$ are the focal elements of $Bel_2 \circ \underline{\theta}_2$.

Now since they are focal elements, the $\omega_1(A_i)$ and $\omega_2(B_j)$ are non-empty. It follows by Theorem 6.10 that the $\omega_1(A_i) \cap \omega_2(B_j)$ are all non-empty. So by Theorem 3.1,

$$m(A) = \sum_{\substack{i,j \\ \omega_1(A_i) \cap \omega_2(B_j) = A}} m'_1(\omega_1(A_i)) m'_2(\omega_2(B_j)) = \sum_{\substack{i,j \\ \omega_1(A_i) \cap \omega_2(B_j) = A}} m_1(A_i) m_2(B_j) .$$

By Lemma 7.7, the sets $\omega_1(A_i) \cap \omega_2(B_j)$ are all distinct, so the lemma follows. ■

Proof of Theorem 7.7. Let $\omega_1 : 2^{\Theta_1} \to 2^{\Omega}$ and $\omega_2 : 2^{\Theta_2} \to 2^{\Omega}$ denote the refinings, and let m_1, m_2, and m denote the basic probability assignments for $S_1 | 2^{\Theta_1}$, $S_2 | 2^{\Theta_2}$, and $S_1 \oplus S_2$, respectively. Then by Lemma 7.8,

$$((S_1 \oplus S_2) | 2^{\Theta_1})(A) = (S_1 \oplus S_2)(\omega_1(A)) = \sum_{B \subset \omega_1(A)} m(B)$$

$$= \sum_{\substack{A_1 \subset \Theta_1, A_2 \subset \Theta_2 \\ \omega_1(A_1) \cap \omega_2(A_2) \subset \omega_1(A)}} m_1(A_1) m_2(A_2)$$

for all $A \subset \Theta_1$. But by Lemma 7.7, $\omega_1(A_1) \cap \omega_2(A_2) \subset \omega_1(A)$ if and only if $A_2 = \emptyset$ or $A_1 \subset A$. So

$$((S_1 \oplus S_2)|2^{\Theta_1})(A) = \sum_{A_1 \subset A} m_1(A_1) \sum_{\substack{A_2 \subset \Theta_2 \\ A_2 \neq \emptyset}} m_2(A_2)$$

$$= \sum_{A_1 \subset A} m_1(A_1)$$

$$= (S_1|2^{\Theta_1})(A) . \blacksquare$$

LEMMA 7.9. *Suppose* Θ_1 *and* Θ_2 *are independent frames of discernment, and let* $\omega_1 : 2^{\Theta_1} \to 2^{\Theta_1 \otimes \Theta_2}$ *and* $\omega_2 : 2^{\Theta_2} \to 2^{\Theta_1 \otimes \Theta_2}$ *denote the refinings to the minimal refinement. Suppose* $A \subset \Theta_1 \otimes \Theta_2$. *If there are no subsets* $B_1 \subset \Theta_1$ *and* $B_2 \subset \Theta_2$ *that satisfy* $A \subset \omega_1(B_1) \cup \omega_2(B_2)$, $A \not\subset \omega_1(B_1)$, *and* $A \not\subset \omega_2(B_2)$, *then* A *is of the form* $\omega_1(A_1) \cap \omega_2(A_2)$ *for some* $A_1 \subset \Theta_1$ *and* $A_2 \subset \Theta_2$.

Proof of Lemma 7.9. Let us prove the contrapositive. In other words, let us assume that A is not of the form $\omega_1(A_1) \cap \omega_2(A_2)$, and then establish the existence of a pair B_1, B_2 satisfying the stated conditions.

By Theorem 6.3, $A \subset \omega_1(\bar{\theta}_1(A)) \cap \omega_2(\bar{\theta}_2(A))$. So if A is not of the form $\omega_1(A_1) \cap \omega_2(A_2)$, there must exist an element θ in $\omega_1(\bar{\theta}_1(A)) \cap \omega_2(\bar{\theta}_2(A))$ that is not in A. Set $B_1 = \overline{\bar{\theta}_1(\{\theta\})}$ and $B_2 = \overline{\bar{\theta}_2(\{\theta\})}$.

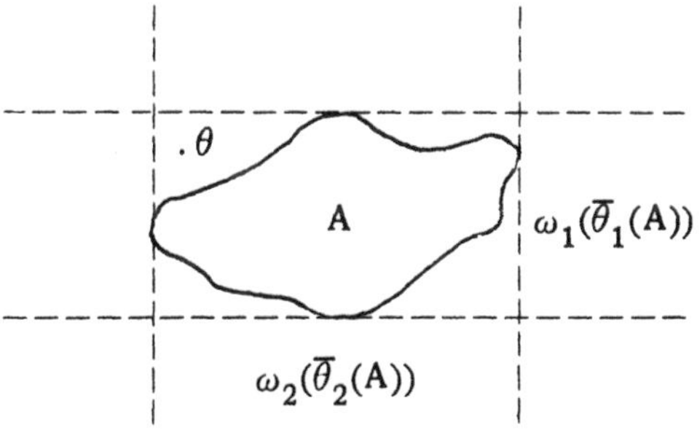

Figure 7.2.

Let us show that $A \not\subset \omega_1(B_1)$. Suppose, indeed, that $A \subset \omega_1(B_1)$. Then by Theorem 6.2, $\bar{\theta}_1(A) \subset B_1$, or $\bar{\theta}_1(A) \subset \overline{\bar{\theta}_1(\{\theta\})}$, or $\bar{\theta}_1(A) \cap \bar{\theta}_1(\{\theta\}) = \emptyset$. But since $\theta \in \omega_1(\bar{\theta}_1(A))$, we know by Theorem 6.3 that $\bar{\theta}_1(\{\theta\}) \subset \bar{\theta}_1(\omega_1(\bar{\theta}_1(A))) = \bar{\theta}_1(A)$, whence $\bar{\theta}_1(A) \cap \bar{\theta}_1(\{\theta\}) = \bar{\theta}_1(\{\theta\})$. So we are led to the false conclusion that $\bar{\theta}_1(\{\theta\}) = \emptyset$. And so really, $A \not\subset \omega_1(B_1)$. Similarly, $A \not\subset \omega_2(B_2)$.

Notice that

$$\overline{\omega_1(B_1) \cup \omega_2(B_2)} = \omega_1(\bar{B}_1) \cap \omega_2(\bar{B}_2)$$

$$= \omega_1(\bar{\theta}_1(\{\theta\})) \cap \omega_2(\bar{\theta}_2(\{\theta\})) .$$

Since $\bar{\theta}_1(\{\theta\})$ and $\bar{\theta}_2(\{\theta\})$ are themselves singletons, it follows from Theorem 6.4 and Lemma 6.2 that this subset is itself a singleton. By Theorem 6.3 it contains $\{\theta\}$; hence it must equal $\{\theta\}$. So

$$\omega_1(B_1) \cup \omega_2(B_2) = \overline{\{\theta\}} \supset A .$$

So B_1 and B_2 satisfy all the required conditions. ∎

Proof of Theorem 7.8.

(i) Let us first prove the equivalence of (1) and (2). It is immediate from the definition that (2) implies (1). In order to prove that (1) implies (2), notice that (1) means that

$$S|2^{\Theta_1 \otimes \Theta_2} = S_1 \oplus S_2 ,$$

where S_1 and S_2 are support functions over $\Theta_1 \otimes \Theta_2$, S_1 is carried by Θ_1, and S_2 is carried by Θ_2. It follows by Theorem 7.7 that

$$S|2^{\Theta_i} = (S|2^{\Theta_1 \otimes \Theta_2})|2^{\Theta_i}$$

$$= S_i|2^{\Theta_i} ,$$

so that S_i is indeed the vacuous extension to $\Theta_1 \otimes \Theta_2$ of $S|2^{\Theta_i}$, for $i = 1, 2$.

(ii) We must also prove the equivalence of (2) and (3). To this end, it is convenient to restate these assertions in terms of the refinings and reductions. Let Bel denote $S|2^{\Theta_1 \otimes \Theta_2}$, let P^* denote its upper probability function, let $\omega_1 : 2^{\Theta_1} \to 2^{\Theta_1 \otimes \Theta_2}$ and $\omega_2 : 2^{\Theta_2} \to 2^{\Theta_1 \otimes \Theta_2}$ be the refinings, and let $\underline{\theta}_1$ and $\underline{\theta}_2$ be the corresponding inner reductions. Then (2) and (3) become:

(2′) $\mathrm{Bel} = (\mathrm{Bel} \circ \omega_1 \circ \underline{\theta}_1) \oplus (\mathrm{Bel} \circ \omega_2 \circ \underline{\theta}_2)$.

(3′) $$\mathrm{Bel}(\omega_1(A_1) \cap \omega_2(A_2)) = \mathrm{Bel}(\omega_1(A_1))\,\mathrm{Bel}(\omega_2(A_2)) \tag{7.7}$$

and

$$P^*(\omega_1(A_1) \cap \omega_2(A_2)) = P^*(\omega_1(A_1))\,P^*(\omega_2(A_2)) \tag{7.8}$$

for all $A_1 \subset \Theta_1$ and $A_2 \subset \Theta_2$.

(iii) Now let us show that (2′) implies (3′). If we let m, m_1 and m_2 denote the basic probability assignments for Bel, $\mathrm{Bel} \circ \omega_1$ and $\mathrm{Bel} \circ \omega_2$, then by Lemmas 7.7 and 7.8, (2′) implies that

$$\begin{aligned}
\mathrm{Bel}(\omega_1(A_1) \cap \omega_2(A_2)) &= \sum_{B \subset \omega_1(A_1) \cap \omega_2(A_2)} m(B) \\
&= \sum_{\substack{B_1 \subset \Theta_1, B_2 \subset \Theta_2 \\ \omega_1(B_1) \cap \omega_2(B_2) \subset \omega_1(A_1) \cap \omega_2(A_2)}} m_1(B_1)\, m_2(B_2) \\
&= \left(\sum_{B_1 \subset A_1} m_1(B_1) \right) \left(\sum_{B_2 \subset A_2} m_2(B_2) \right) \\
&= \mathrm{Bel}(\omega_1(A_1))\,\mathrm{Bel}(\omega_2(A_2))
\end{aligned}$$

whenever $A_1 \subset \Theta_1$ and $A_2 \subset \Theta_2$. Similarly, using (2.8), we find that

$$P^*(\omega_1(A_1)\cap\omega_2(A_2)) = \sum_{\substack{B\subset\Theta_1\otimes\Theta_2\\ B\cap(\omega_1(A_1)\cap\omega_2(A_2))\neq\emptyset}} m(B)$$

$$= \sum_{\substack{B_1\subset\Theta_1, B_2\subset\Theta_2\\ (\omega_1(B_1)\cap\omega_2(B_2))\cap(\omega_1(A_1)\cap\omega_2(A_2))\neq\emptyset}} m_1(B_1)m_2(B_2)$$

$$= \left(\sum_{\substack{B_1\subset\Theta_1\\ A_1\cap B_1\neq\emptyset}} m_1(B_1)\right)\left(\sum_{\substack{B_2\subset\Theta_2\\ A_2\cap B_2\neq\emptyset}} m_2(B_2)\right)$$

$$= P^*(\omega_1(A_1))\,P^*(\omega_2(A_2))$$

whenever $A_1\subset\Theta_1$ and $A_2\subset\Theta_2$. So (2′) implies (3′).

(iv) Now let us assume (3′) and derive (2′). Notice that (7.8) is equivalent to

$$1 - \mathrm{Bel}(\overline{\omega_1(A_1)\cap\omega_2(A_2)}) = (1-\mathrm{Bel}(\overline{\omega_1(A_1)}))\,(1-\mathrm{Bel}(\overline{\omega_2(A_2)}))\ ,$$

or

$$\mathrm{Bel}(\omega_1(\bar{A}_1)\cup\omega_2(\bar{A}_2)) = \mathrm{Bel}(\omega_1(\bar{A}_1)) + \mathrm{Bel}(\omega_2(\bar{A}_2)) - \mathrm{Bel}(\omega_1(\bar{A}_1))\,\mathrm{Bel}(\omega_2(\bar{A}_2))$$

for all $A_1\subset\Theta_1$ and $A_2\subset\Theta_2$. If we change $\bar{A}_1$ to A_1 and $\bar{A}_2$ to A_2 and substitute (7.7) in the result, we obtain

$$\mathrm{Bel}(\omega_1(A_1)\cup\omega_2(A_2)) = \mathrm{Bel}(\omega_1(A_1)) + \mathrm{Bel}(\omega_2(A_2)) - \mathrm{Bel}(\omega_1(A_1)\cap\omega_2(A_2))$$

for all $A_1\subset\Theta_1$ and $A_2\subset\Theta_2$.

In terms of Bel's basic probability assignment m, this relation becomes

$$\sum_{B\subset\omega_1(A_1)\cup\omega_2(A_2)} m(B) = \sum_{\substack{B\subset\Theta_1\otimes\Theta_2\\ B\subset\omega_1(A_1)\text{ or }B\subset\omega_2(A_2)}} m(B)\ .$$

We may conclude that $m(B) = 0$ whenever there are $B_1 \subset \Theta_1$ and $B_2 \subset \Theta_2$ such that $B \subset \omega_1(B_1) \cup \omega_2(B_2)$ but B is contained in neither $\omega_1(B_1)$ nor $\omega_2(B_2)$. So by Lemma 7.9, the focal elements of Bel are all the form $\omega_1(B_1) \cap \omega_2(B_2)$ for various $B_1 \subset \Theta_1$ and $B_2 \subset \Theta_2$.

Using this fact and (1) of Lemma 7.7, (7.7) becomes

$$\sum_{B_1 \subset A_1} \sum_{B_2 \subset A_2} m(\omega_1(B_1) \cap \omega_2(B_2)) = \sum_{B_1 \subset A_1} \sum_{B_2 \subset A_2} m_1(B_1) m_2(B_2)$$

for all $A_1 \subset \Theta_1$ and $A_2 \subset \Theta_2$. It follows, by induction on $\max(|B_1|, |B_2|)$, that

$$m(\omega_1(B_1) \cap \omega_2(B_2)) = m_1(B_1) m_2(B_2)$$

for all $B_1 \subset \Theta_1$ and $B_2 \subset \Theta_2$. So by Lemma 7.8, Bel has the same basic probability assignment as, and hence is equal to, $(\mathrm{Bel} \circ \underline{\theta}_1) \oplus (\mathrm{Bel} \circ \underline{\theta}_2)$. ∎

LEMMA 7.10. *Suppose* Bel_1 *and* Bel_2 *are belief functions over* Θ *and* $\mathrm{Bel}_1 \oplus \mathrm{Bel}_2$ *exists. Let* P_1^*, P_2^* *and* P *denote the upper probability functions for* Bel_1, Bel_2 *and* $\mathrm{Bel}_1 \oplus \mathrm{Bel}_2$, *and let* m_1, m_2 *and* m *denote the basic probability assignments. Then*

$$P^*(A) = \frac{\sum_{B \subset \Theta} m_1(B) P_2^*(A \cap B)}{\sum_{B \subset \Theta} m_1(B) P_2^*(B)}$$

for all $A \subset \Theta$.

Proof of Lemma 7.10. By (2.8) and Theorem 3.1,

$$P^*(A) = \sum_{\substack{D \subset \Theta \\ A \cap D \neq \emptyset}} m(D)$$

$$= \frac{\displaystyle\sum_{\substack{B,C \\ A \cap B \cap C \neq \emptyset}} m_1(B)\, m_2(C)}{\displaystyle\sum_{\substack{B,C \\ B \cap C \neq \emptyset}} m_1(B)\, m_2(C)}$$

$$= \frac{\displaystyle\sum_{B \subset \Theta} m_1(B) \sum_{\substack{C \subset \Theta \\ A \cap B \cap C \neq \emptyset}} m_2(C)}{\displaystyle\sum_{B \subset \Theta} m_1(B) \sum_{\substack{C \subset \Theta \\ B \cap C \neq \emptyset}} m_2(C)}$$

$$= \frac{\displaystyle\sum_{B \subset \Theta} m_1(B)\, P_2^*(A \cap B)}{\displaystyle\sum_{B \subset \Theta} m_1(B)\, P_2^*(B)} . \quad \blacksquare$$

Proof of Theorem 7.9.

(i) First let us assume that Θ_1 and Θ_2 are cognitively independent with respect to S and then prove that (7.5) holds for all A discerned by Θ_1 and B discerned by Θ_2.

Suppose A is discerned by Θ_1, and let $S_1 : 2^{\Omega} \to [0,1]$ be the simple support function focused on A with $S_1(A) = 1$. Then S_1 is carried by Θ_1.

If $Pl(A) = 0$, then

$$Pl(A \cap B) = 0 = Pl(A)Pl(B)$$

for all B discerned by Θ_2. If $Pl(A) > 0$, then by Theorem 3.6, $S \oplus S_1$ exists and its plausibility function, say Pl', is given by

$$Pl'(B) = \frac{Pl(A \cap B)}{Pl(A)} \tag{7.9}$$

for all $B \subset \Omega$. If B is discerned by Θ_2, then the hypothesis that Θ_1 and Θ_2 are cognitively independent implies that $Pl'(B) = Pl(B)$, so (7.9) yields (7.5).

(ii) Now let us assume that (7.5) holds for all A discerned by Θ_1 and B discerned by Θ_2 and then prove that Θ_1 and Θ_2 are cognitively independent with respect to S.

Suppose S_1 is a support function over Ω that is carried by Θ_1, and denote its basic probability assignment by m_1. Then all of S_1's focal elements are discerned by Θ_1. So if $S \oplus S_1$ exists, then by Lemma 7.10 its plausibility function Pl' satisfies

$$Pl'(B) = \frac{\sum_{A \subset \Omega} m_1(A)\,Pl(A \cap B)}{\sum_{A \subset \Omega} m_1(A)\,Pl(A)} = \frac{\sum_{\substack{A \subset \Omega \\ m_1(A) \neq \emptyset}} m_1(A)\,Pl(A)\,Pl(B)}{\sum_{\substack{A \subset \Omega \\ m_1(A) \neq \emptyset}} m_1(A)\,Pl(A)}$$

$$= Pl(B)$$

for all B discerned by Θ_2. So $(S \oplus S_1)|2^{\Theta_2} = S|2^{\Theta_2}$. Since S_1 was an arbitrary support function carried by Θ_1, it follows that Θ_1 and Θ_2 are cognitively independent. ■

CHAPTER 8. THE DISCERNMENT OF EVIDENCE

> After all, important fresh evidence is a two-edged thing, and may possibly cut in a very different direction to what Lestrade imagines.
>
> SHERLOCK HOLMES

The last two chapters have brought us to a global point of view. As we now see it, a body of evidence simultaneously affects a whole family of compatible frames of discernment, determining a support function on each member of that family. This point of view is a natural outgrowth of our earlier ideas, yet it provides a new perspective on some of those ideas. In particular, it provides a new perspective on Dempster's rule of combination and on the idea of weights of evidence.

In the introduction to Chapter 3, I mentioned that one prerequisite for using Dempster's rule of combination for combining two bodies of evidence is that one's frame of discernment should discern the relevant interaction of those bodies of evidence. This observation became more intelligible when we learned about refinements of frames of discernment, but we still need to know how to tell when a frame does discern the interaction of two bodies of evidence. This question is addressed in §1 and §2 of this chapter.

The possibility of refinement – especially indefinite refinement – poses obvious problems for the idea that degrees of support are based on weights of evidence. Indeed, the weights of evidence associated with a separable support function retain their intuitive meaning only so long as the separable support function is thought to capture the whole impact of the evidence; they seem to lose that meaning when the separable support function is thought to be merely the restriction of a more complicated support function over a refinement. In §3 and §4, I offer some tentative suggestions about how the notion of weights of evidence might be preserved under the prospect of indefinite refinement.

§1. Families of Compatible Support Functions

If a given body of evidence $\mathfrak{E}$ determines a support function over every frame Θ in a family $\mathfrak{F}$, then it determines a *family of compatible support functions* $\{S_{\mathfrak{E}}^{\Theta}\}_{\Theta\epsilon\mathfrak{F}}$ – a collection of belief functions which are support functions and which are compatible in the sense explained in §3 of Chapter 6. The complexity of the structure of this family of support functions will depend on whether or not $\mathfrak{E}$ *affects* $\mathfrak{F}$ *sharply*.

DEFINITION. The evidence $\mathfrak{E}$ *affects* $\mathfrak{F}$ *sharply* if there exists a frame $\Theta\epsilon\mathfrak{F}$ that carries $S_{\mathfrak{E}}^{\Omega}$ for every $\Omega\epsilon\mathfrak{F}$ that is a refinement of Θ. Such a frame Θ is said to *exhaust the impact of* $\mathfrak{E}$ *on* $\mathfrak{F}$.

When Θ exhausts the impact of $\mathfrak{E}$ on $\mathfrak{F}$, $S_{\mathfrak{E}}^{\Theta}$ determines the whole family $\{S_{\mathfrak{E}}^{\Omega}\}_{\Omega\epsilon\mathfrak{F}}$. For if Θ carries the support function over every refinement of Θ, then the support function over any given frame $\Omega\epsilon\mathfrak{F}$ will be the restriction to Ω of $S_{\mathfrak{E}}^{\Theta}$'s vacuous extension to $\Theta\otimes\Omega$. But if $\mathfrak{F}$ does not affect $\mathfrak{E}$ sharply, then no single frame in $\mathfrak{F}$ will be fine enough to exhaust all the detail of the effect of $\mathfrak{E}$; it will always be possible to discern further effects of the evidence by further refinement of one's frame.

It is obviously easiest to deal with examples in which the evidence affects the family of frames sharply; in such cases one may choose a frame that exhausts the impact of the evidence on the family and then carry out one's analysis with no fear of overlooking any subtleties. And such examples often occur. They are typical whenever the frames of discernment and the evidence are highly idealized, as they are in the case of statistical evidence (see Chapter 11), and perhaps even in the case of scientific evidence in general (see Chapter 12).

Yet one can imagine examples where the evidence does not affect one's family of frames sharply, and one can even argue that such examples are typical in our everyday experience. After all, an item of evidence will scarcely seem to have any substance unless it makes manifold contacts with the questions that frame our perception of the world.

Consider, for example, the evidence that was supposed to cast suspicion on left-handers in the burglary of the sweetshop. One can hardly weave a convincing tale about how evidence supported the proposition that the thief was a left-hander without showing that evidence to have supported other propositions along the way, and hence I found it necessary to suggest that the evidence also indicated certain specifics about how the safe was opened. Accepting this suggestion, we might refine the frame $\Theta = \{LI, LO, RI, RO\}$ of Example 4.1, obtaining a frame Ω that distinguishes between different ways in which the safe might have been opened, between different causes of various things observed around the safe, between different ways in which the safe might have been manufactured, etc. And we might suppose that with respect to Ω Holmes' evidence can be decomposed into homogeneous components, each determining a simple support function over Ω. Yet it is difficult to specify what a focus of one of these simple support functions might be without again admitting that the evidence can support such a subset only if it also supports other propositions, yet more remote and detailed and hence not discerned by Θ. In short, it might be difficult to put a limit on the specification of detail in the evidence.

On the other hand, we must remember that a body of evidence is not, in the present theory, a collection of material objects or other "real evidence" that can be subjected to repeated examination.* Rather, it is

*The term "real evidence" is used to describe tangible objects introduced into evidence in a court and displayed for examination by the judge and jury. Such evidence must be identified by sworn testimony. See the article "Evidence" in the 1973 printing of the *Encyclopaedia Britannica*.

a segment of our experience and natural knowledge, the result of those observations and investigations we have actually performed. And one might well argue that our experience and natural knowledge is essentially finite and limited. We can always demand further details of a witness, but as Pierre Duhem pointed out, his memory and perception will become more and more hazy as we do so.* Eventually we reach a point where everything fades into confusion, and that confusion presumably signals a frame exhausting the detail of the witness's evidence.

But whether or not we can always posit the existence of a frame that exhausts the detail of the evidence, there will be many cases where we will not actually want to specify such a frame; our experience is often far too rich to be exhausted by any framework describable with a practical amount of time and effort. So there is in any case adequate practical reason to study the situation where the evidence does not affect our family of frames sharply – where one must deal with frames that do not exhaust the impact of the evidence.

§2. Discerning the Interaction of Evidence

We cannot properly combine several bodies of evidence unless we take into account all the relevant ways in which those bodies of evidence interact. Indeed, it is a commonplace that by selecting particular inferences from one body of evidence and combining them with particular inferences from another body of evidence, one can arrive at conclusions quite at variance with those that would emerge from a more thorough combination of the evidence.

This feature of the combination of evidence is reflected in our theory by the fact that Dempster's rule of combination may give inaccurate results when it is applied in too coarse a frame of discernment. Suppose, indeed, that S_1 and S_2 are support functions over a frame Ω, that they are

*See §V of Chapter IV of the second part of Duhem's *La Theorie physique: Son objet et sa structure.*

based on distinct bodies of evidence, and that Θ is a coarsening of Ω. Then if we apply Dempster's rule in the frame Ω, we obtain the support function

$$(S_1 \oplus S_2)|2^{\Theta}$$

as our support function over Θ, but if we apply it in the frame Θ, we obtain the support function

$$(S_1|2^{\Theta}) \oplus (S_2|2^{\Theta}) .$$

And these two support functions may differ.

EXAMPLE 8.1. *The Burglary Again.* Let us denote by Ω the frame of discernment that we used to analyze Example 4.1:

$$\Omega = \{LI, LO, RI, RO\} .$$

Set $\Theta = \{\theta_1, \theta_2\}$, and let $\omega : 2^{\Theta} \to 2^{\Omega}$ be the refining given by $\omega(\{\theta_1\}) = \{LI\}$ and $\omega(\{\theta_2\}) = \{LO, RI, RO\}$. Thus the coarsening Θ distinguishes only whether the left-handed insider was or was not the culprit.

One of the bodies of evidence in Example 4.1 cast suspicion on left-handers and thus produced a simple support function S_1 over Ω that focused on

$$A = \{LI, LO\} ;$$

the other cast suspicion on insiders and thus produced a simple support function S_2 over Ω that focused on

$$B = \{LI, RI\} .$$

Neither S_1 nor S_2 record any support for either $\{LI\}$ or $\{LO, RI, RO\}$; hence $S_1|2^{\Theta}$, $S_2|2^{\Theta}$ and $(S_1|2^{\Theta}) \oplus (S_2|2^{\Theta})$ are all vacuous. But $(S_1 \oplus S_2)|2^{\Theta}$ is not vacuous. For $S_1 \oplus S_2$ provides the degree of support

$$S_1(A)\, S_2(B) > 0$$

for $A \cap B = \{LI\}$, and its restriction to 2^Θ will thus provide that same degree of support for $\{\theta_1\}$. ∎

The frame of discernment Θ failed to discern the support for $\{\theta_1\}$ in this example because it could discern S_1's support for A only as support for $\bar{\theta}(A) = \Theta$, it could discern S_2's support for B only as support for $\bar{\theta}(B) = \Theta$, and

$$\{\theta_1\} = \bar{\theta}(A \cap B) \neq \bar{\theta}(A) \cap \bar{\theta}(B) = \Theta .$$

Now there will always be subsets A and B of Ω such that $\bar{\theta}(A \cap B) \neq \bar{\theta}(A) \cap \bar{\theta}(B)$. But it seems intuitively clear that the coarsening Θ of Ω can be appropriate for combining two bodies of evidence $\mathcal{E}_1$ and $\mathcal{E}_2$ only if

$$\bar{\theta}(A \cap B) = \bar{\theta}(A) \cap \bar{\theta}(B) \tag{8.1}$$

whenever A is a subset of Ω that is exactly supported by $\mathcal{E}_1$ and B is a subset of Ω that is exactly supported by $\mathcal{E}_2$. And this condition is indeed sufficient to ensure that the combination over Θ agrees with the combination over Ω:

THEOREM 8.1. *Suppose* S_1 *and* S_2 *are support functions over a frame* Ω, $\bar{\theta}: 2^\Omega \to 2^\Theta$ *is an outer reduction,* $S_1 \oplus S_2$ *exists, and (8.1) holds whenever* A *is a focal element of* S_1 *and* B *is a focal element of* S_2. *Then*

$$(S_1|2^\Theta) \oplus (S_2|2^\Theta) = (S_1 \oplus S_2)|2^\Theta .$$

Thus we are led to the following definition:

DEFINITION. Suppose S_1 and S_2 are support functions over a frame Ω and $\bar{\theta}: 2^\Omega \to 2^\Theta$ is an outer reduction. Then Θ is said to *discern the relevant interaction of* S_1 *and* S_2 if (8.1) holds whenever A is a focal element of S_1 and B is a focal element of S_2.

Let us record two intuitively appropriate consequences of this definition.

THEOREM 8.2. *Suppose the support functions* S_1 *and* S_2 *over a frame* Ω *are carried by a coarsening* Θ. *Then* Θ *discerns the relevant interaction of* S_1 *and* S_2.

THEOREM 8.3. *Suppose* S_1 *and* S_2 *are two support functions over a frame* Ω, Θ *is a coarsening of* Ω, *and* Θ_0 *is a coarsening of* Θ. *If* Θ_0 *discerns the relevant interaction of* S_1 *and* S_2, *then* Θ_0 *discerns the relevant interaction of* $S_1|2^\Theta$ *and* $S_2|2^\Theta$.

Theorem 8.2 is reassuring; it tells us that Θ discerns the relevant interaction of S_1 and S_2 if it discerns everything about S_1 and S_2. But there are certainly cases where Θ discerns the relevant interaction of S_1 and S_2 even though it does not discern all the detail of either. In particular, Θ will discern the relevant interaction of S_1 and S_2 if the detail of S_1 that it fails to discern is "independent" of the detail of S_2 that it fails to discern:

THEOREM 8.4. *Suppose* Θ, Θ_1, *and* Θ_2 *are independent coarsenings of a frame* Ω, *suppose* S_1 *is a support function over* $\Theta\otimes\Theta_1\otimes\Theta_2$ *that is carried by* $\Theta\otimes\Theta_1$, *and suppose* S_2 *is a support function over* $\Theta\otimes\Theta_1\otimes\Theta_2$ *that is carried by* $\Theta\otimes\Theta_2$. *Then* Θ *discerns the relevant interaction of* S_1 *and* S_2.

To say that a coarsening Θ of a frame Ω discerns the relevant interaction of the support function $S_1 : 2^\Omega \to [0, 1]$ determined by a body of evidence $\mathcal{E}_1$ and the support function $S_2 : 2^\Omega \to [0,1]$ determined by a body of evidence $\mathcal{E}_2$ is to say that Θ discerns as much of the relevant interaction of $\mathcal{E}_1$ and $\mathcal{E}_2$ as is discerned by Ω. But we will not be willing to use Θ to combine $\mathcal{E}_1$ and $\mathcal{E}_2$ unless we can hope that Θ discerns *all* the relevant interaction of $\mathcal{E}_1$ and $\mathcal{E}_2$.

DEFINITION. Suppose $\mathcal{F}$ is a family of compatible frames, $\{S_1^\Omega\}_{\Omega\epsilon\mathcal{F}}$ is the family of support functions determined by a body of evidence $\mathcal{E}_1$, and $\{S_2^\Omega\}_{\Omega\epsilon\mathcal{F}}$ is the family of support functions determined by the body of evidence $\mathcal{E}_2$. Then a particular frame Θ in $\mathcal{F}$ is said to *discern the relevant interaction of* $\mathcal{E}_1$ *and* $\mathcal{E}_2$ if

$$\bar{\theta}(A \cap B) = \bar{\theta}(A) \cap \bar{\theta}(B)$$

whenever Ω is a refinement of Θ, $\bar{\theta}: 2^\Omega \to 2^\Theta$ is the outer reduction, A is a focal element of S_1^Ω, and B is a focal element of S_2^Ω.

We see from Theorem 8.2 that a frame in a family $\mathcal{F}$ will certainly discern the relevant interaction of $\mathcal{E}_1$ and $\mathcal{E}_2$ if the frame exhausts the impact of both $\mathcal{E}_1$ and $\mathcal{E}_2$ on $\mathcal{F}$. So in the case where $\mathcal{E}_1$ and $\mathcal{E}_2$ both affect $\mathcal{F}$ sharply, we have a theoretical method, at least, for deciding whether a particular frame $\Theta\epsilon\mathcal{F}$ is appropriate for combining $\mathcal{E}_1$ and $\mathcal{E}_2$: we identify a frame $\Omega\epsilon\mathcal{F}$ that exhausts the impact of both $\mathcal{E}_1$ and $\mathcal{E}_2$ on $\mathcal{F}$ and then check whether $\bar{\theta}: 2^\Omega \to 2^\Theta$ obeys (8.1) whenever A is a focal element of $S_{\mathcal{E}_1}^\Omega$ and B is focal element of $S_{\mathcal{E}_2}^\Omega$.

But there will certainly be cases where we cannot hope to find a frame that exhausts the impact of both $\mathcal{E}_1$ and $\mathcal{E}_2$ on $\mathcal{F}$ – either because $\mathcal{E}_1$ and $\mathcal{E}_2$ do not affect $\mathcal{F}$ sharply and no such frame exists, or perhaps more realistically, because such a frame would be too hopelessly complicated to be within our grasp in practice. In such cases, we may be able to appeal to the following theorem.

THEOREM 8.5. *Suppose $\mathcal{F}$ is a family of compatible frames, $\{S_1^{\Theta'}\}_{\Theta'\epsilon\mathcal{F}}$ is the family of support functions determined by the body of evidence $\mathcal{E}_1$, and $\{S_2^{\Theta'}\}_{\Theta'\epsilon\mathcal{F}}$ is the family of support functions determined by the body of evidence $\mathcal{E}_2$. Suppose $\Theta\epsilon\mathcal{F}$. And suppose that for every refinement Ω of Θ, there exist Θ_1 and Θ_2 in $\mathcal{F}$ such that*

(1) *Θ, Θ_1, and Θ_2 are independent,*

(2) *$\Theta \otimes \Theta_1 \otimes \Theta_2$ is a refinement of Ω,*

(3) *$S_1^{\Theta \otimes \Theta_1 \otimes \Theta_2}$ is carried by $\Theta \otimes \Theta_1$,*

(4) *$S_2^{\Theta \otimes \Theta_1 \otimes \Theta_2}$ is carried by $\Theta \otimes \Theta_2$.*

Then Θ discerns the relevant interaction of $\mathcal{E}_1$ and $\mathcal{E}_2$.

Although the conditions of this theorem seem complicated, there are situations where we can feel confident that they are met.

EXAMPLE 8.2. *The Evidence Against Insiders.* In our first discussion of the burglary, we noted that Holmes' evidence against left-handers was derived from an inspection of the opened safe, together of course with his general knowledge about safes. We never learned the exact nature of his evidence against insiders, but we tacitly assumed that that evidence was entirely distinct from the evidence about the safe. In order to be definite, let us now suppose that the evidence against insiders was derived from an inspection of the doors of the store and information about how the locking of those doors was managed.

Holmes has, then, two bodies of evidence $\mathcal{E}_1$ and $\mathcal{E}_2$. The evidence $\mathcal{E}_1$ has the simple effect of supporting {LI, LO} when it is analyzed with respect to the frame

$$\Theta = \{\text{LI, LO, RI, RO}\}$$

but would presumably exhibit more complicated effects as Θ were refined so as to touch on details about how the safe was opened, what kind of safe it was, how such safes work, etc. Similarly, the evidence $\mathcal{E}_2$ has the simple effect of supporting {LI, RI} when it is analyzed with respect to Θ, but would exhibit more complicated effects as Θ were refined so as to touch on details about the appearance of the doors after the burglary, how the doors were locked, etc.

Intuitively, Holmes is entitled to use the frame Θ to combine $\mathcal{E}_1$ and $\mathcal{E}_2$ if he is confident that the details about the safe and the details about the doors are independent. More precisely, he must be confident that no matter how far Θ is refined, the effect of $\mathcal{E}_1$ that is thereby discerned will be exhausted by a frame $\Theta \otimes \Theta_1$ and the effect of $\mathcal{E}_2$ that is thereby discerned will be exhausted by a frame $\Theta \otimes \Theta_2$, where Θ_1 is concerned only with details about the safe and about left-handers, Θ_2 is concerned only with details about the doors and insiders, and Θ, Θ_1, and Θ_2 are independent. Fortunately, Holmes may well be able to acquire such confidence from a general appraisal of $\mathcal{E}_1$ and $\mathcal{E}_2$, without spelling out all their endless detail. ■

When we say that a frame discerns the *relevant* interaction of two bodies of evidence, we mean, of course, *relevant to that frame*. If there is any detail of the evidence that the frame does not discern, then there will be interaction that it does not discern. But that interaction may be relevant only to a finer frame.

EXAMPLE 8.3. *Irrelevant Interaction*. Set $\Omega = \{a, b, c, d\}$, $\Theta = \{\theta_1, \theta_2, \theta_3\}$, and $\Theta_0 = \{\alpha, \delta\}$. Define a refining $\omega : 2^\Theta \to 2^\Omega$ by

$$\omega(\{\theta_1\}) = \{a\},$$

$$\omega(\{\theta_2\}) = \{b, c\},$$

$$\omega(\{\theta_3\}) = \{d\}.$$

And define a refining $\omega_0 : 2^{\Theta_0} \to 2^\Omega$ by

$$\omega_0(\{\alpha\}) = \{a, b, c\},$$

$$\omega_0(\{\delta\}) = \{d\}.$$

Then the outer reductions $\bar{\theta} : 2^\Omega \to 2^\Theta$ and $\bar{\theta}_0 : 2^\Omega \to 2^{\Theta_0}$ obey

$$\begin{aligned} \bar{\theta}(\{a\}) &= \{\theta_1\}, & \bar{\theta}_0(\{a\}) &= \{a\}, \\ \bar{\theta}(\{a,b\}) &= \{\theta_1,\theta_2\}, & \bar{\theta}_0(\{a,b\}) &= \{a\} \\ \bar{\theta}(\{a,c\}) &= \{\theta_1,\theta_2\}, & \bar{\theta}_0(\{a,c\}) &= \{a\}, \\ \bar{\theta}(\{a,b,c\}) &= \{\theta_1,\theta_2\}, & \bar{\theta}_0(\{a,b,c\}) &= \{a\}. \end{aligned}$$

Now let $S_1 : 2^\Omega \to [0,1]$ be the simple support function focused on $A = \{a,b\}$ with $S_1(A) = \frac{1}{2}$, say, and let $S_2 : 2^\Omega \to [0,1]$ be the simple support function focused on $B = \{a,c\}$ with $S_2(B) = \frac{1}{2}$. Then

$$\bar{\theta}_0(A \cap B) = \{a\} = \bar{\theta}_0(A) \cap \bar{\theta}_0(B),$$

so Θ_0 discerns the relevant interaction of S_1 and S_2.

There certainly is interaction between S_1 and S_2 that Θ_0 does not discern. For S_1's support for A interacts with S_2's support for B to produce support for $A \cap B = \{a\}$. But this interaction is irrelevant to Θ_0, which cannot distinguish $A \cap B$ from either A or B.

Notice, though, that this interaction *is* relevant to Θ, which also fails to discern it. Indeed,

$$\bar{\theta}(A \cap B) = \{\theta_1\} \neq \{\theta_1,\theta_2\} = \bar{\theta}(A) \cap \bar{\theta}(B);$$

Θ fails to discern all the relevant interaction of S_1 and S_2, even though it is finer than Θ_0. ■

§3. Discerning Weights of Evidence

We have just seen that Dempster's rule of combination can be valid even when some details of the evidence are ignored. In this section and the next, we ask whether weights of evidence can be sensibly defined and discussed when some of the details of the evidence are ignored. As we will see, this is an open mathematical question.

The role played by weights of evidence in our theory began to recede in importance as soon as we introduced the notion of refinement. When we developed the idea of weights of evidence in Chapters 4 and 5, we tacitly assumed that our frame Θ exhausted the impact of the evidence; the evidence was supposed to be decomposable into homogeneous components, each having the precise and sole effect of supporting a subset of Θ. Under these circumstances, it was natural to talk about weights of evidence, and we could pay as much attention to the assessment w over Θ as to the corresponding separable support function S. But when we introduced the notion of a refinement and supposed that the evidence might appear more complicated when viewed from a refinement of Θ than when viewed from Θ itself, we lost the notion that there would necessarily be weights of evidence associated with subsets of Θ; instead, we might have a non-separable support function over Θ, with weights of evidence associated only with subsets of some refinement.

And now that we have admitted the possibility that there might be no refinement of our frame Θ that completely exhausts the impact of the evidence, or at least that such a refinement might be beyond the practical scope of our attention, our weights of evidence seem to have disappeared altogether. If the support function over each frame that we consider is merely the restriction of a more complicated support function over a refinement, then as we consider finer and finer frames we may never encounter a support function that is separable. And even if we do, its separability will seem only accidental; the associated assessment will have no intuitive meaningfulness if further refinement is known to reveal further complication.

Yet the intuitive attraction of the notion of weights of evidence remains. It would be pleasant if we could associate weights of evidence or something like them with every support function S, whether or not S is separable, and whether or not we examine the support functions over refinings of S's frame. After all, we can concede that that frame may fail to discern some aspects of the evidence and still hope to speak of the "weight" of the evidence it does discern.

Suppose, then, that we consider a support function $S: 2^{\Theta} \rightarrow [0,1]$ in isolation. We suppose that S arose as the restriction of some separable support function over some refinement Ω of Θ, but we deny knowing any detail about this separable support function or even about Ω. What can we say about the "weights of evidence" underlying S?

The natural reaction to this question is to consider the set

$$\mathfrak{W}_S = \{w \mid w \text{ is an assessment of evidence over some refinement of } \Theta, \text{ and the separable support function } S_w \text{ determined by } w \text{ satisfies } S_w|2^\Theta = S.\}$$

and ask whether there are features common to all the elements in it – i.e., to all the assessments that might have produced S. This is a reasonably well-posed mathematical problem, but unfortunately not one that I have been able to solve. In the rest of this chapter I record a few facts and ideas about $\mathfrak{W}_S$, in the hope that some readers will be able to carry the problem further.

One fact to note about $\mathfrak{W}_S$ is that it always includes assessments some of whose weights are arbitrarily large.

THEOREM 8.6. *Suppose* $S: 2^\Theta \to [0,1]$ *is a support function and* N *is a positive number. Then there exists a refining* $\omega: 2^\Theta \to 2^\Omega$, *an assessment* $w: 2^\Omega \to [0,\infty]$ *in* $\mathfrak{W}_S$, *and a subset* $A \subset \Omega$ *such that* $w(A) = N$.

This is, of course, merely one aspect of the fact that Θ does not discern everything; there can easily be many weights on subsets of Ω that have no impact at all on degrees of support for subsets of Θ.

A second fact to note is that $\mathfrak{W}_S$ will include assessments that focus no weights at all on any of the propositions discerned by Θ.

THEOREM 8.7. *Suppose* $S: 2^\Theta \to [0,1]$ *is a support function. Then there exists a refining* $\omega: 2^\Theta \to 2^\Omega$ *and an assessment* $w: 2^\Omega \to [0,\infty]$ *in* $\mathfrak{W}_S$ *such that* $w(\omega(A)) = 0$ *for all* $A \subset \Theta$.

The reason for this is obvious: in order to arrange for support for a subset A of Θ, one need not focus a weight precisely on $\omega(A)$; a slightly smaller subset of Ω will do.

Theorem 8.6 suggests that we should look for the weights of evidence that are minimal, in some sense, for determining S. But Theorem 8.7 makes it clear that we should not try to minimize the weights focused exactly on subsets corresponding to the subsets of Θ. An obvious alternative is to minimize the values of the impingement function.

§4. If the Weight-of-Conflict Conjecture is True

Suppose $S: 2^{\Theta} \to [0,1]$ is a support function, and set

$$\mathcal{V}_S = \{V \mid V \text{ is an impingement function over some refinement of } \Theta, \text{ and the separable support function } S_V \text{ determined by } V \text{ satisfies } S_V|2^{\Theta} = S.\}$$

For each $V \in \mathcal{V}_S$, let Ω_V denote the refinement of Θ over which V is defined, let $\omega_V : 2^{\Theta} \to 2^{\Omega_V}$ denote the refining, and let W_V denote the weight of internal conflict for S_V. Set

$$\underline{W}_S = \inf_{V \in \mathcal{V}_S} W_V , \tag{8.2}$$

and for each $A \subset \Theta$ set

$$\underline{V}_S(A) = \inf_{V \in \mathcal{V}_S} V(\omega_V(A)) . \tag{8.3}$$

The quantity $\underline{W}_S$ may be called the *weight of conflict discerned by* Θ, and the function $\underline{V}_S : 2^{\Theta} \to [0,\infty]$ defined by (8.3) may be called the *discerned impingement function for* Θ.

THEOREM 8.8. *Suppose* S *is a separable support function,* W_S *is its weight of internal conflict, and* V_S *is its impingement function. If the weight-of-conflict conjecture is true, then* $W_S = \underline{W}_S$ *and* $V_S = \underline{V}_S$.

In other words, if the weight-of-conflict conjecture is true, then (8.2) and (8.3) succeed to some degree in extending the notion of weights of evidence from separable support functions to support functions in general.

It has been my experience that the quantities W_V and $V(\omega_V(A))$ tend to increase as the element V of $\mathcal{O}_S$ and its refining ω_V are made more complicated. (Example 5.3 provides an instance of this.) Hence I am inclined to believe that the infima in (8.2) and (8.3) can always be attained with relatively simple refinements. But I have no insight into how such refinements might be constructed or how the quantities (8.2) and (8.3) might be calculated.

We might ask whether the discerned impingement function $\underline{V}_S$ fully determines S – or whether two support functions S_1 and S_2 might sometimes have the same discerned impingement function. It would be intuitively pleasing if different support functions always produced different discerned impingement functions, but I have no evidence that they do.

§5. Mathematical Appendix

Proof of Theorem 8.1. Let $m_1, m_2, m, m'_1, m'_2, m'$ and m'' denote the basic probability assignments for $S_1, S_2, S_1 \oplus S_2$, $S_1|2^\Theta, S_2|2^\Theta$, $(S_1 \oplus S_2)|2^\Theta$ and $(S_1|2^\Theta) \oplus (S_2|2^\Theta)$, respectively. Then by Theorems 8.1 and 6.9,

$$m'(A) = \sum_{\substack{B \subset \Omega \\ A = \bar{\theta}(B)}} m(B)$$

$$= \sum_{\substack{B \subset \Omega \\ A = \bar{\theta}(B)}} \frac{\displaystyle\sum_{\substack{B_1, B_2 \subset \Omega \\ B_1 \cap B_2 = B}} m_1(B_1)\, m_2(B_2)}{\displaystyle\sum_{\substack{B_1, B_2 \subset \Omega \\ B_1 \cap B_2 \neq \emptyset}} m_1(B_1)\, m_2(B_2)}$$

$$= \frac{\displaystyle\sum_{\substack{B_1,B_2 \subset \Omega \\ \bar{\theta}(B_1 \cap B_2) = A}} m_1(B_1)\, m_2(B_2)}{\displaystyle\sum_{\substack{B_1,B_2 \subset \Omega \\ B_1 \cap B_2 \neq \emptyset}} m_1(B_1)\, m_2(B_2)} ,$$

and

$$m''(A) = \frac{\displaystyle\sum_{\substack{A_1,A_2 \subset \Theta \\ A_1 \cap A_2 = A}} m'_1(A_1)\, m'_2(A_2)}{\displaystyle\sum_{\substack{A_1,A_2 \subset \Theta \\ A_1 \cap A_2 \neq \emptyset}} m'_1(A_1)\, m'_2(A_2)}$$

$$= \frac{\displaystyle\sum_{\substack{A_1,A_2 \subset \Theta \\ A_1 \cap A_2 = A}} \left(\sum_{\substack{B_1 \subset \Omega \\ A_1 = \bar{\theta}(B_1)}} m_1(B_1) \right) \left(\sum_{\substack{B_2 \subset \Omega \\ A_2 = \bar{\theta}(B_2)}} m_2(B_2) \right)}{\displaystyle\sum_{\substack{A_1,A_2 \subset \Theta \\ A_1 \cap A_2 \neq \emptyset}} \left(\sum_{\substack{B_1 \subset \Omega \\ A_1 = \bar{\theta}(B_1)}} m_1(B_1) \right) \left(\sum_{\substack{B_2 \subset \Omega \\ A_2 = \bar{\theta}(B_2)}} m_2(B_2) \right)}$$

$$= \frac{\displaystyle\sum_{\substack{B_1,B_2 \subset \Omega \\ \bar{\theta}(B_1) \cap \bar{\theta}(B_2) = A}} m_1(B_1)\, m_2(B_2)}{\displaystyle\sum_{\substack{B_1,B_2 \subset \Omega \\ \bar{\theta}(B_1) \cap \bar{\theta}(B_2) \neq \emptyset}} m_1(B_1)\, m_2(B_2)}$$

for all non-empty $A \subset \Theta$. It is evident from these formulae that $m' = m''$ if $\bar{\theta}(A \cap B) = \bar{\theta}(A) \cap \bar{\theta}(B)$ whenever A is a focal element of S_1 and B is a focal element of S_2. ∎

Proof of Theorem 8.2. Let $\omega : 2^\Theta \to 2^\Omega$ be the refining in question. Since S_1 and S_2 are carried by Θ, their focal elements are all of the form $\omega(C)$ for $C \subset \Theta$. But by Theorem 6.3, $\bar{\theta} \circ \omega$ is the identity. So if $A = \omega(C)$ is a focal element of S_1 and $B = \omega(D)$ is a focal element of S_2, then

$$\begin{aligned}\bar{\theta}(A \cap B) &= \bar{\theta}(\omega(C) \cap \omega(D)) = \bar{\theta}(\omega(C \cap D)) \\ &= C \cap D = \bar{\theta}(\omega(C)) \cap \bar{\theta}(\omega(D)) \\ &= \bar{\theta}(A) \cap \bar{\theta}(B) . \blacksquare\end{aligned}$$

LEMMA 8.1. *Suppose $\bar{\theta} : 2^\Omega \to 2^\Theta$ is an outer reduction. Then*

$$\bar{\theta}(A \cap B) \subset \bar{\theta}(A) \cap \bar{\theta}(B)$$

for all $A, B \subset \Omega$.

Proof of Lemma 8.1. By (7) of Theorem 6.3, $\bar{\theta}(A \cap B) \subset \bar{\theta}(A)$ and $\bar{\theta}(A \cap B) \subset \bar{\theta}(B)$. ■

LEMMA 8.2. *Suppose $\omega_0 : 2^{\Theta_0} \to 2^\Theta$ and $\omega : 2^\Theta \to 2^\Omega$ are refinings, and denote their outer reductions by $\bar{\theta}_0$ and $\bar{\theta}$, respectively. Then $\bar{\theta}_0 \circ \bar{\theta}$ is the outer reduction for the refining $\omega \circ \omega_0$.*

Proof of Lemma 8.2.

$$\begin{aligned}(\bar{\theta}_0 \circ \bar{\theta})(A) &= \bar{\theta}_0(\{\theta \epsilon \Theta | \omega(\{\theta\}) \cap A \neq \emptyset\}) \\ &= \{\theta_0 \epsilon \Theta_0 | \omega_0(\{\theta_0\}) \cap \{\theta \epsilon \Theta | \omega(\{\theta\}) \cap A \neq \emptyset\} \neq \emptyset\} \\ &= \{\theta_0 \epsilon \Theta_0 | \text{ there exists } \theta \epsilon \omega_0(\{\theta_0\}) \text{ such that } \omega(\{\theta\}) \cap A \neq \emptyset\} \\ &\equiv \left\{\theta_0 \epsilon \Theta_0 \middle| \left(\bigcup_{\theta \epsilon \omega_0(\{\theta_0\})} \omega(\{\theta\})\right) \cap A \neq \emptyset\right\} \\ &= \{\theta_0 \epsilon \Theta_0 | \omega(\omega_0(\{\theta_0\})) \cap A \neq \emptyset\} \\ &= \{\theta_0 \epsilon \Theta_0 | (\omega \circ \omega_0)(\{\theta_0\}) \cap A \neq \emptyset\} . \blacksquare\end{aligned}$$

Proof of Theorem 8.3. Let $\overline{\theta}: 2^{\Omega} \to 2^{\Theta}$ and $\overline{\theta}_0 : 2^{\Theta} \to 2^{\Theta_0}$ denote the outer reductions in question. If C and D are any subsets of Ω, then

$$\overline{\theta}(C \cap D) \subset \overline{\theta}(C) \cap \overline{\theta}(D)$$

by Lemma 8.1, and hence

$$\overline{\theta}_0(\overline{\theta}(C \cap D)) \subset \overline{\theta}_0(\overline{\theta}(C) \cap \overline{\theta}(D)) \tag{8.5}$$

by (7) of Theorem 6.3. And by applying Lemma 8.1 to $\overline{\theta}_0$, we obtain

$$\overline{\theta}_0(\overline{\theta}(C) \cap \overline{\theta}(D)) \subset \overline{\theta}_0(\overline{\theta}(C)) \cap \overline{\theta}_0(\overline{\theta}(D)) . \tag{8.6}$$

Now suppose $A \subset \Theta$ is a focal element of $S_1 | 2^{\Theta}$ and $B \subset \Theta$ is a focal element of $S_2 | 2^{\Theta}$. Then we must show that $\overline{\theta}_0(A \cap B) = \overline{\theta}_0(A) \cap \overline{\theta}_0(B)$. But it is clear from Theorem 6.9 that $A = \overline{\theta}(C)$ and $B = \overline{\theta}(D)$, where C is a focal element of S_1 and D is a focal element of S_2. Furthermore, Θ_0 discerns all the relevant interaction between S_1 and S_2 that is discerned by Ω, and $\overline{\theta}_0 \circ \overline{\theta}$ is the coarsening from Ω to Θ_0. So

$$\overline{\theta}_0(\overline{\theta}(C)) \cap \overline{\theta}_0(\overline{\theta}(D)) = \overline{\theta}_0(\overline{\theta}(C \cap D)) . \tag{8.7}$$

Comparing (8.5), (8.6), and (8.7), we find that

$$\overline{\theta}_0(\overline{\theta}(C) \cap \overline{\theta}(D)) = \overline{\theta}_0(\overline{\theta}(C)) \cap \overline{\theta}_0(\overline{\theta}(D)) ,$$

or $\overline{\theta}_0(A \cap B) = \overline{\theta}_0(A) \cap \overline{\theta}_0(B)$. ■

LEMMA 8.3. *Suppose Θ_1, Θ_2, and Θ_3 are independent frames, and denote $\Omega = \Theta_1 \otimes \Theta_2 \otimes \Theta_3$. Let $\omega_1 : 2^{\Theta_1} \to 2^{\Omega}$, $\omega_1^0 : 2^{\Theta_1} \to 2^{\Theta_1 \otimes \Theta_2}$, $\omega_2 : 2^{\Theta_2} \to 2^{\Omega}$, $\omega_2^0 : 2^{\Theta_2} \to 2^{\Theta_1 \otimes \Theta_2}$, and $\omega : 2^{\Theta_1 \otimes \Theta_2} \to 2^{\Omega}$ be the refinings between the indicated frames. Then*

$$\omega_1(\{\theta_1\}) \cap \omega_2(\{\theta_2\}) \neq \emptyset$$

whenever $\theta_1 \in \Theta_1$ *and* $\theta_2 \in \Theta_2$. *And if* $A \subset \Theta_1 \otimes \Theta_2$, *then there exists a subset* A' *of the Cartesian product* $\Theta_1 \times \Theta_2$ *such that*

$$\omega(A) = \bigcup_{(\theta_1,\theta_2)\in A'} (\omega_1(\{\theta_1\}) \cap \omega_2(\{\theta_2\})) . \tag{8.8}$$

Proof of Lemma 8.3. The fact that $\omega_1(\{\theta_1\}) \cap \omega_2(\{\theta_2\}) \neq \emptyset$ is an immediate consequence of (4) of Theorem 6.10.

By (2) of Theorem 6.4, there exists $A' \subset \Theta_1 \times \Theta_2$ such that

$$A = \bigcup_{(\theta_1,\theta_2)\in A'} (\omega_1^0(\{\theta_1\}) \cap \omega_2^0(\{\theta_2\})) .$$

Since $\omega_1 = \omega \circ \omega_1^0$ and $\omega_2 = \omega \circ \omega_2^0$, (8.8) follows. ■

Proof of Theorem 8.4. Denote $\Omega = \Theta \otimes \Theta_1 \otimes \Theta_2$, and let $\omega : 2^{\Theta} \to 2^{\Omega}$, $\omega_1 : 2^{\Theta_1} \to 2^{\Omega}$, $\omega_2 : 2^{\Theta_2} \to 2^{\Omega}$, $\omega_{01} : 2^{\Theta \otimes \Theta_1} \to 2^{\Omega}$ and $\omega_{02} : 2^{\Theta \otimes \Theta_2} \to 2^{\Omega}$ be the refinings between the indicated frames. Let A be a focal element of S_1, and let B be a focal element of S_2.

Since S_1 is carried by $\Theta \otimes \Theta_1$, A will be of the form $\omega_{01}(C)$, where $C \subset \Theta \otimes \Theta_1$. Similarly, B will be of the form $\omega_{02}(D)$, where $D \subset \Theta \otimes \Theta_2$. So by Lemma 8.3, there exists a subset C' of $\Theta \times \Theta_1$ and a subset D' of $\Theta \times \Theta_2$ such that

$$A = \omega(C) = \bigcup_{(\theta,\theta_1)\in C'} \omega(\{\theta\}) \cap \omega_1(\{\theta_1\})$$

and

$$B = \omega(D) = \bigcup_{(\theta,\theta_2)\in D'} \omega(\{\theta\}) \cap \omega_2(\{\theta_2\}) .$$

Using the fact that each $\omega(\{\theta\}) \cap \omega(\{\theta_1\})$ is non-empty, together with the fact that the $\omega(\{\theta\})$ are disjoint, we find that

$$\bar{\theta}(A) = \left\{\theta' \epsilon \Theta \,|\, \omega(\{\theta'\}) \cap \left(\bigcup_{(\theta,\theta_1)\epsilon C'} \omega(\{\theta\}) \cap \omega_1(\{\theta_1\}) \right) \neq \emptyset \right\}$$

$$= \{\theta \epsilon \Theta \,|\, \text{There exists } \theta_1 \epsilon \Theta_1 \text{ such that } (\theta,\theta_1) \epsilon C'.\},$$

where $\bar{\theta}: 2^{\Omega} \to 2^{\Theta}$ is the outer reduction. Similarly,

$$\bar{\theta}(B) = \{\theta \epsilon \Theta \,|\, \text{There exists } \theta_2 \epsilon \Theta_2 \text{ such that } (\theta,\theta_2) \epsilon D'.\}.$$

And

$$\bar{\theta}(A \cap B) = \{\theta \epsilon \Theta \,|\, \text{There exist } \theta_1 \epsilon \Theta_1 \text{ and } \theta_2 \epsilon \Theta_2 \text{ such that } (\theta,\theta_1) \epsilon C' \text{ and } (\theta,\theta_2) \epsilon D'.\}$$
$$= \bar{\theta}(A) \cap \bar{\theta}(B). \blacksquare$$

Proof of Theorem 8.5. Suppose Ω is an arbitrary refinement of Θ, and suppose Θ_1 and Θ_2 satisfy the conditions listed. Then by Theorem 8.4, Θ discerns the relevant interaction between $S_1^{\Theta\otimes\Theta_1\otimes\Theta_2}$ and $S_2^{\Theta\otimes\Theta_1\otimes\Theta_2}$. It follows by Theorem 8.3 that Θ discerns the relevant interaction between S_1^{Ω} and S_2^{Ω}. And since Ω was arbitrary, it follows that Θ discerns all the relevant interaction between $\mathcal{E}_1$ and $\mathcal{E}_2$. ■

LEMMA 8.4. *Suppose* $S_0: 2^{\Theta} \to [0,1]$ *is a separable support function with assessment* w_0, *and* $\omega: 2^{\Theta} \to 2^{\Omega}$ *is a refining. Let* S *denote the vacuous extension of* S_0 *to* Ω. *Then* S *is separable, and its assessment* w *is given by*

$$w(A) = \begin{cases} w_0(B) & \text{if } A = \omega(B), \\ 0 & \text{if } A \neq \omega(B) \text{ for any } B \subset \Theta. \end{cases}$$

In other words, $w(\omega(B)) = w_0(B)$ *for all* $B \subset \Theta$, *and* $w(A) = 0$ *whenever* A *is not discerned by* Θ.

Furthermore, the weight of internal conflict for S *is equal to the weight of internal conflict for* S_0. *And if* V *and* V_0 *are the impingement functions for* S *and* S_0, *then* $V(\omega(B)) = V_0(B)$ *for all* $B \subset \Theta$.

Proof of Lemma 8.4. Let $\mathcal{C}_0$ denote the core of S_0. Then the function w defined above satisfies $w(\omega(\mathcal{C}_0)) = \infty$ and $w(A) = 0$ for all $A \not\subset \omega(\mathcal{C}_0)$. Hence w is an assessment of evidence with core $\omega(\mathcal{C}_0)$. According to Theorem 5.4, its commonality function Q′ is given by

$$\begin{aligned} Q'(A) &= K \exp\left(-\sum_{\substack{B\subset\Omega \\ A\not\subset B}} w(B)\right) \\ &= K \exp\left(-\sum_{\substack{C\subset\Theta \\ A\not\subset\omega(C)}} w_0(C)\right) \\ &= K \exp\left(-\sum_{\substack{C\subset\Theta \\ \bar{\theta}(A)\not\subset C}} w_0(C)\right) \end{aligned}$$

for all non-empty $A \subset \Omega$.

According to Theorem 7.4, the commonality function for S, which we may denote by Q, is given by $Q = Q_0 \circ \bar{\theta}$, or

$$Q(A) = Q_0(\bar{\theta}(A)) = K_0 \exp\left(-\sum_{\substack{C\subset\Theta \\ \bar{\theta}(A)\not\subset C}} w_0(C)\right),$$

for some constant K_0. But $K = K_0$ by (2.6), and hence $Q = Q'$. So w is the assessment of evidence for S.

By Theorem 5.6, $W_S = \log K$ and $W_{S_0} = \log K_0$. So we have also shown that $W_S = W_{S_0}$. And

$$\begin{aligned} V(\omega(B)) = \sum_{\substack{C\subset\Omega \\ \omega(B)\not\subset C}} w(B) &= \sum_{\substack{D\subset\Theta \\ \omega(B)\not\subset\omega(D)}} w_0(D) = \sum_{\substack{D\subset\Theta \\ B\not\subset D}} w_0(D) \\ &= V(B) \end{aligned}$$

for all $B \subset \Theta$. ■

LEMMA 8.5. *If* Θ *is a member of a family* $\mathcal{F}$ *of compatible frames of discernment, then there exists a frame* Θ_0 *of* $\mathcal{F}$ *that contains more than one element and is independent of* Θ.

Proof of Lemma 8.5. By condition (5) of the definition of a family of compatible frames, there exists, for any given element $\theta \in \Theta$, a refining that maps θ to a set of two elements. It follows that there exists a refining $\omega : 2^{\Theta} \to 2^{\Omega}$ such that $\omega(\{\theta\})$ has at least two elements for every $\theta \in \Theta$. Given such a refining ω, form a set $A \subset \Omega$ by including exactly one of the elements in each $\omega(\{\theta\})$. Then $A \neq \emptyset$ and $A \neq \Omega$, and for each $\theta \in \Theta$, A and $\bar{A}$ have non-empty intersections with $\omega(\{\theta\})$.

By condition (4) of the definition of a family of compatible frames, there exists a frame $\Theta_0 = \{\theta_1, \theta_2\}$ in $\mathcal{F}$ and a refining $\omega_0 : 2^{\Theta_0} \to 2^{\Omega}$ such that $\omega_0(\{\theta_1\}) = A$ and $\omega_0(\{\theta_2\}) = \bar{A}$.

It follows from (4) of Theorem 6.11 that Θ and Θ_0 are independent. ∎

Proof of Theorem 8.6. By the definition of a support function, there exists a refinement Ω_0 of Θ and a separable support function S'_0 over Ω_0 such that $S = S'_0|2^{\Theta}$.

By Lemma 8.5, there exists a frame Ω_1 that has more than one element and is compatible with Ω_0 but independent of it. Set $\Omega = \Omega_0 \otimes \Omega_1$.

Let S_0 denote the natural extension of S'_0 to Ω, and let S_1 be a simple support function focused on a proper non-empty subset A_1 of Ω that is discerned by Ω_1, with $S_1(A_1) = 1 - e^{-N}$. Then S_0 is carried by Ω_0 and S_1 is carried by Ω_1. By Theorem 7.7, $S_0 \oplus S_1$ exists and

$$(S_0 \oplus S_1)|2^{\Omega_0} = S_0|2^{\Omega_0} = S'_0 ,$$

whence

$$(S_0 \oplus S_1)|2^{\Theta} = S'_0|2^{\Theta} = S .$$

By Lemma 8.4, S_0 is separable. Suppose $S_0 = S_2 \oplus \cdots \oplus S_n$ is the canonical decomposition of S_0 into simple support functions, with each

S_i focused on $A_i \subset \Omega$ and based on the weight of evidence w_i, $i=2,\cdots,n$. Then Lemma 8.1 assures us that $A_2,\cdots,A_n$ are all discerned by Ω_0. If we let $\mathcal{C}_0$ denote the core of S_0, then we know that $\mathcal{C}_0$ is also discerned by Ω_0.

Notice that $\mathcal{C}_0$ is also the core of $S_0 \oplus S_1$.

Set $w_1 = N$. Then the assessment w for $S_0 \oplus S_1$ is given by

$$w(A) = \sum_{\substack{i \\ A_i \cap \mathcal{C}_0 = A}} w_i$$

for all $A \subset \mathcal{C}_0$. In particular,

$$w(A_1 \cap \mathcal{C}_0) = \sum_{\substack{i \\ A_i \cap \mathcal{C}_0 = A_1 \cap \mathcal{C}_0}} w_i$$

$$= N + \sum_{\substack{i \\ 2 \leq i \leq n \\ A_i \cap \mathcal{C}_0 = A_1 \cap \mathcal{C}_0}} w_i \,.$$

But since $A_i \subset \mathcal{C}_0$ for $i=2,\cdots,n$, the requirement that $A_i \cap \mathcal{C}_0 = A_1 \cap \mathcal{C}_0$ reduces to $A_i = A_1 \cap \mathcal{C}_0$, or $\Omega \cap A_i = A_1 \cap \mathcal{C}_0$. Since $A_1 \neq \Omega$, it follows from (3) of Lemma 7.7 that this condition cannot be met. So $w(A_1 \cap \mathcal{C}_0) = N$. ∎

Proof of Theorem 8.7. Construct Ω_0, S'_0, Ω_1, Ω, S_0, S_1, and $S_0 \oplus S_1$ as in the preceding proof. But set $N = \infty$. Then it will still be true that $(S_0 \oplus S_1)|2^\Theta = S$. But now the core $\mathcal{C}$ of $S_0 \oplus S_1$ will be equal to $\mathcal{C}_0 \cap A_1$, where $\mathcal{C}_0$ is the core of S_0.

Suppose $A \subset \Omega$ is discerned by Θ. Then $A = A \cap \Omega$, and by (1) of Lemma 7.7, $A \subset \mathcal{C}_0 \cap A_1$ if and only if both $A \subset \mathcal{C}_0$ and $\Omega \subset A_1$. But A_1 is a proper subset of Ω by assumption, whence $\Omega \not\subset A_1$. So $A \not\subset \mathcal{C}_0 \cap A_1$, and hence the assessment w for $S_0 \oplus S_1$ must satisfy $w(A) = 0$. ∎

Proof of Theorem 8.8. Let Θ denote the frame for S. Suppose $\omega : 2^\Theta \to 2^\Omega$ is a refining, and suppose S' is a separable support function over Ω such that $S'|2^\Theta = S$. Let Q' and V' denote the commonality and impingement functions for S'.

Let Q denote the commonality function for the vacuous extension of S to Ω. Lemma 8.4 tells us that Q is separable, that $W_Q = W_S$, and that the impingement function V for Q is related to V_S by $V(\omega(A)) = V_S(A)$ for all $A \subset \Theta$.

By Theorem 7.5,

$$Q(A) \geq Q'(A)$$

for all $A \subset \Omega$. So if the weight-of-conflict conjecture is true, then

$$W_Q \leq W_{Q'},$$

and

$$V(A) \leq V'(A)$$

for all $A \subset \Omega$. (See Theorem 5.8.)

Since $V \in \mathcal{V}_S$ and V' is an arbitrary second element of $\mathcal{V}_S$, it follows that

$$\underline{W}_S = \inf_{V' \in \mathcal{V}_S} W_{V'} = W_V = W_S$$

and

$$\underline{V}_S(A) = \inf_{V' \in \mathcal{V}_S} V'(\omega_{V'}(A)) = V(\omega(A)) = V_S(A)$$

for all $A \subset \Theta$. ∎

CHAPTER 9. QUASI SUPPORT FUNCTIONS

> ··· when the defect of data is supplied by hypothesis, the solutions will, in general, vary with the nature of the hypotheses assumed ···
>
> GEORGE BOOLE (1815-1864)

In this chapter, we look more closely at those belief functions that are not support functions – the belief functions I call *quasi support functions*. We begin by learning how quasi support functions can be represented as limits of sequences of support functions, and we then turn to the implications of this representation for an important subclass of the quasi support functions, the non-trivial Bayesian belief functions.

As we see in §1 and §2, the limit of a sequence of separable support functions will be a quasi support function provided that the weight of evidence impugning each possibility in the frame of discernment tends to infinity. Intuitively, this means that a quasi support function corresponds to the specification of infinite contradictory weights of evidence. Such weights of evidence cannot be combined directly, but the limiting process allows them to balance each other and to produce degrees of belief that depend on their differences. In point of fact, of course, it makes no sense to talk about the difference of two infinite quantities; so quasi support functions are best thought of as idealizations that cannot represent actual evidence.

In §§3-5, we turn our attention to degrees of belief based on chances and to the belief functions modeled after the paradigm of chance, the Bayesian belief functions. In §3, I briefly relate the notion of infinite contradictory weights of evidence to our usual understanding of the hypothetical character of chances. In §4, I show that a Bayesian belief function indicates infinite evidence *favoring* each possibility in its frame of discernment. And in §5, I use our general theory to gain a new perspective on the assimilation of evidence by a Bayesian.

§1. Infinite Contradictory Evidence

In §5 of Chapter 4, we considered the case of an assessment $w : 2^{\Theta} \to [0, \infty]$ that assigned positive weights of evidence to only two disjoint subsets A and B. The result was a separable support function S with the three focal elements A, B, and Θ, and with basic probability numbers

$$m(A) = S(A) = \frac{e^{w(A)} - 1}{e^{w(A)} + e^{w(B)} - 1} ,$$

$$m(B) = S(B) = \frac{e^{w(B)} - 1}{e^{w(A)} + e^{w(B)} - 1} ,$$

and

$$m(\Theta) = \frac{1}{e^{w(A)} + e^{w(B)} - 1} .$$

Of course, either $w(A)$ or $w(B)$ may be infinite, in which case $m(\Theta)$ and either $m(B)$ or $m(A)$ will be zero. But it is presupposed, both by these formulae and by the definition of an assessment of evidence, that not both $w(A)$ and $w(B)$ are infinite. One cannot combine certainties in two contradictory propositions.

Suppose, however, that we begin with finite values for $w(A)$ and $w(B)$ and then let those values tend to infinity while holding the difference $w(B) - w(A)$ equal to a constant Δ, or at least letting it tend to the finite value Δ. Then the basic probability numbers will also tend to definite limits:

$$m(A) = \frac{e^{w(A)} - 1}{(1+e^{\Delta})e^{w(A)} - 1} \longrightarrow \frac{1}{1 + e^{\Delta}} \equiv m_{\infty}(A) ,$$

$$m(B) = \frac{e^{\Delta}e^{w(A)} - 1}{(1+e^{\Delta})e^{w(A)} - 1} \longrightarrow \frac{e^{\Delta}}{1+e^{\Delta}} \equiv m_{\infty}(B) ,$$

and

$$m(\Theta) = \frac{1}{(1+e^{\Delta})e^{w(A)} - 1} \longrightarrow 0 \equiv m_{\infty}(\Theta) .$$

And the function $m_{\infty} : 2^{\Theta} \to [0,1]$ to which m tends satisfies

$$\sum_{C \subset \Theta} m_{\infty}(C) = m_{\infty}(A) + m_{\infty}(B) = 1 ,$$

and thus qualifies as a basic probability assignment. This basic probability assignment seems to correspond to a situation where the contradictory weights $w(A)$ and $w(B)$ are both infinite, yet differ precisely by the finite amount $\Delta = w(B) - w(A)$.

One way of describing the situation is to say that every point of Θ is impugned by an infinite weight of evidence. All of $\bar{A}$ is impugned by the infinite weight $w(A)$, all of $\bar{B}$ is impugned by the infinite weight $w(B)$, and $\bar{A} \cup \bar{B} = \Theta$.

The belief function $Bel_{\infty} : 2^{\Theta} \to [0,1]$ determined by m_{∞} is simply described:

$$Bel_{\infty}(C) = \begin{cases} 0 & \text{if } C \text{ contains neither } A \text{ nor } B \\ \dfrac{1}{1+e^{\Delta}} & \text{if } C \text{ contains } A \text{ but not } B \\ \dfrac{e^{\Delta}}{1+e^{\Delta}} & \text{if } C \text{ contains } B \text{ but not } A \\ 1 & \text{if } C \text{ contains both } A \text{ and } B . \end{cases}$$

And it has two noteworthy properties. First, the degrees of belief for A and B are additive:

$$\mathrm{Bel}_\infty(A) + \mathrm{Bel}_\infty(B) = \frac{1}{1+e^{\Delta}} + \frac{e^{\Delta}}{1+e^{\Delta}}$$

$$= 1$$

$$= \mathrm{Bel}_\infty(A \cup B) \ .$$

And secondly, Bel_∞ is not a support function: its core $A \cup B$ is not a focal element.

It is difficult to imagine a belief function such as Bel_∞ being useful for the representation of actual evidence. Even when the accumulation of evidence is so extreme as to make infinite contradictory weights of evidence a reasonable idealization, it still remains awkward to imagine these infinite weights differing by some finite amount. Yet this same vision of infinite but delicately balanced weights of evidence arises for all belief functions that are not support functions.

§2. The Class of Quasi Support Functions

The preceding example illustrated two general facts. First, any belief function that is not itself a support function can be obtained either directly as the limit of a sequence of separable support functions or indirectly as the restriction of such a limit. And secondly, whenever such a sequence does converge to a belief function that is not a support function, the weight of evidence impugning each possibility in the frame of discernment becomes infinite. In this section, I state these two facts as theorems and use them to justify the epithet *quasi support function* to name those belief functions that are not support functions.

Since we are concerned only with functions on a finite power set 2^Θ, there are no subtleties involved in the notion of a limit of a sequence of functions; a sequence $f_1, f_2 \cdots$ of functions on 2^Θ is said to tend to the limit f if

$$\lim_{i \to \infty} f_i(A) = f(A)$$

for all $A \subset \Theta$. Furthermore, the introduction of limits does not lead to any class of functions larger than the class of belief functions:

> THEOREM 9.1. *If a sequence of belief functions has a limit, the limit is a belief function.*

But as I just asserted, limits do lead out of the class of support functions to the whole class of belief functions:

> THEOREM 9.2. *If the belief function* $\mathrm{Bel}: 2^\Theta \to [0,1]$ *is not a support function, then there exists a refinement* Ω *of* Θ *and a sequence* $S_1, S_2, \cdots$ *of separable support functions over* Ω *such that* $\mathrm{Bel} = \left(\lim_{i\to\infty} S_i\right)|2^\Theta$.

(If a sequence $S_1, S_2, \cdots$ of separable support functions over Ω converges, then

$$\left(\lim_{i\to\infty} S_i\right)|2^\Theta = \lim_{i\to\infty} (S_i|2^\Theta) .$$

And the sequence $S_1|2^\Theta, S_2|2^\Theta, \cdots$ is a sequence of support functions over Θ. So to say that Bel is the restriction of the limit of a sequence of separable support functions is to say that Bel is the limit of a sequence of support functions. The latter assertion is the simpler, but the former is more convenient when we are concerned with weights of evidence.)

Whenever a sequence of separable support functions over a frame Θ tends to a limit that is not a separable support function, inspection will show that $V(\{\theta\})$, the total weight of evidence impugning θ, has become infinite for every $\theta \in \Theta$:

> THEOREM 9.3. *Suppose the sequence* $S_1, S_2, \cdots$ *of separable support functions over* Θ *tends to a limit that is not a separable support function, and denote the impingement functions for* $S_1, S_2, \cdots$ *by* $V_1, V_2, \cdots$. *Then*

$$\lim_{i \to \infty} V_i(\{\theta\}) = \infty$$

for every $\theta \in \Theta$.

Because of the dubious nature of such infinite contradictory weights of evidence, it is natural to call a belief function a *quasi support function* whenever it is not a support function but is the limit of a sequence of separable support functions or the restriction of such a limit. In this vocabulary, Theorem 9.2 tells us that every belief function that is not a support function is a quasi support function.

§3. Chances are not Degrees of Support

Any belief function that assigns non-trivially additive degrees of belief to a proposition and its negation must be classed as a quasi support function:

> THEOREM 9.4. *Suppose* Bel *is a belief function over* Θ. *Suppose* $A \subset \Theta$, $\mathrm{Bel}(A) > 0$, $\mathrm{Bel}(\bar{A}) > 0$, *and* $\mathrm{Bel}(A) + \mathrm{Bel}(\bar{A}) = 1$. *Then* Bel *is a quasi support function.*

Thus most Bayesian belief functions cannot qualify as support functions.

> THEOREM 9.5. *A Bayesian belief function* $\mathrm{Bel}: 2^{\Theta} \to [0,1]$ *qualifies as a support function if and only if there exists* $\theta \in \Theta$ *such that* $\mathrm{Bel}(\{\theta\}) = 1$.

And in particular, a Bayesian belief function obtained by adopting chances as one's degrees of belief will not qualify as a support function.

Those who are accustomed to thinking of partial beliefs based on chances as paradigmatic may be startled to see them relegated to a peripheral role and classified among those partial beliefs that cannot arise from actual, finite evidence. But students of statistical inference are quite familiar with the conclusion that a chance cannot be evaluated with less than infinite evidence. To establish a value between zero and one as the chance for a given outcome of an aleatory process, one must obtain the results of an infinite sequence of independent trials of the process, and the proportion of those trials that result in the outcome must converge to a value not equal to zero or one as one proceeds through this sequence. One could ask for no better example of infinite, precisely balanced and unobtainable evidence.

Chances, then, are essentially hypothetical rather than empirical, and can seldom be translated directly into degrees of support. Yet they do play an important role in probable reasoning. As I mentioned in Chapter 1, one can often incorporate chances into a frame of discernment by associating a different chance for some event with each different possibility in the frame, and this can facilitate the evaluation of that event as evidence. This role of chances is taken up again in Chapter 11.

§4. The Bayesian Profusion of Infinite Weights

As Theorem 9.5 tells us, the only Bayesian belief functions that are support functions are of a very trivial type: they concentrate all one's belief on a single point. Most Bayesian belief functions are quasi support functions, and as such they indicate infinite weights of evidence impugning each of the possibilities in their frames of discernment. And we can say even more: in a certain sense, a Bayesian belief function indicates an infinite amount of evidence *in favor of* each possibility in its core.

Recall that an assessment of evidence $w : 2^{\Theta} \to [0, \infty]$ may indicate some evidence in favor of an element $\theta \in \Theta$ even though $w(\{\theta\}) = 0$; all that is required is that there exist $A_1, \cdots, A_n \subset \Theta$ such that $w(A_i) > 0$ for all i and $A_1 \cap \cdots \cap A_n = \{\theta\}$. In general, we would say that there is some evidence favoring θ if and only if

$$\max_{\substack{\mathcal{A} \subset 2^{\Theta} \\ \cap \mathcal{A} = \{\theta\}}} \min_{B \in \mathcal{A}} w(B) > 0 .$$

And we would say that there is infinite evidence favoring θ whenever this same quantity becomes infinite. So the following theorem justifies the assertion that a Bayesian belief function indicates infinite evidence in favor of each element of its core.

THEOREM 9.6. *Suppose* $w_1, w_2, \cdots$ *is a sequence of assessments of evidence with respect to a frame* Θ, *and suppose the corresponding separable support functions converge to a belief function* Bel *with core* $\mathcal{C} \subset \Theta$. *Then* Bel *is Bayesian if and only if*

$$\lim_{i\to\infty} \max_{\substack{\mathcal{A}\subset 2^{\Theta} \\ \cap\mathcal{A}=\{\theta\}}} \min_{B\in\mathcal{A}} w_i(B) = \infty \tag{9.1}$$

for each $\theta \in \mathcal{C}$. *And if* $\omega : 2^{\Theta_0} \to 2^{\Theta}$ *is a refining, then* $\mathrm{Bel}|2^{\Theta_0}$ *is Bayesian if and only if*

$$\lim_{i\to\infty} \max_{\substack{\mathcal{A}\subset 2^{\Theta} \\ \emptyset\neq\cap\mathcal{A}\subset\omega(\{\theta_0\})}} \min_{B\in\mathcal{A}} w_i(B) = \infty \tag{9.2}$$

for every θ_0 *in the core of* $\mathrm{Bel}|2^{\Theta_0}$.

EXAMPLE 9.1. *A Profusion of Infinite Weights*. Set $\Theta = \{a,b,c,d\}$ and consider the effect of focusing infinite but equal weights of evidence on various subsets of Θ as illustrated in Figure 9.1.

Case 1. First, focus equal but infinite weights on the singletons, while focusing no weight on other subsets. We can do this

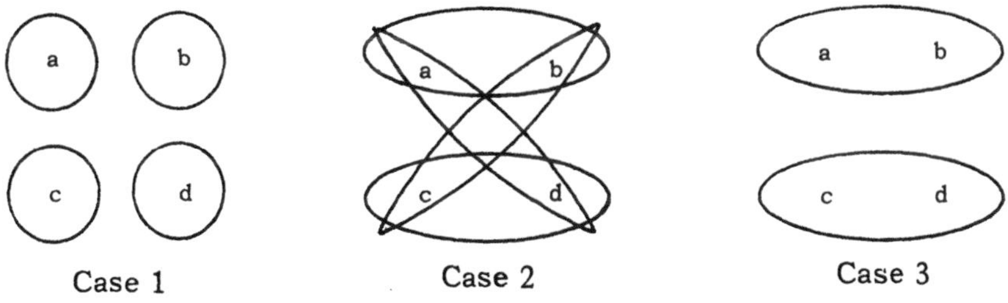

Figure 9.1.

through a limiting process if we define, for each natural number i, an assessment w_i, where

$$w_i(\{a\}) = w_i(\{b\}) = w_i(\{c\}) = w_i(\{d\}) = i$$

and $w_i(A) = 0$ for all other $A \subset \Theta$. The corresponding separable support functions then tend, as i tends to infinity, to the Bayesian belief function Bel given by

$$\mathrm{Bel}(\{a\}) = \mathrm{Bel}(\{b\}) = \mathrm{Bel}(\{c\}) = \mathrm{Bel}(\{d\}) = \frac{1}{4} .$$

(Since Bel is Bayesian, its values for non-singletons are determined by addition.)

Case 2. This time, define w_i by

$$w_i(\{a,b\}) = w_i(\{c,d\}) = w_i(\{a,d\}) = w_i(\{b,c\}) = i$$

and $w_i(A) = 0$ for all other $A \subset \Theta$. The belief function obtained from these assessments as i tends to infinity is the same as that of Case 1.

Case 3. The last example provides a contrast, for it is a quasi support function that is not Bayesian. Define w_i by

$$w_i(\{a,b\}) = w_i(\{c,d\}) = i$$

and $w_i(A) = 0$ for all other $A \subset \Theta$. Then one obtains the belief function with focal elements $\{a,b\}$ and $\{c,d\}$, each having a basic probability number of $\frac{1}{2}$. Though each singleton is impugned by an infinite weight of evidence (the infinite weight focused on $\{a,b\}$ impugns both $\{c\}$ and $\{d\}$, while the infinite weight focused on $\{c,d\}$ impugns both $\{a\}$ and $\{b\}$), there is no evidence favoring any particular singleton. ■

§5. Bayes' Theorem

Precisely because it already indicates an infinite amount of evidence favoring each possibility, a Bayesian belief function obscures many of the features of any actual evidence with which it is combined. This means that the assimilation of new evidence is simplified for anyone beginning with a Bayesian belief function.

One way to describe this simplification is to point out that the combination of a Bayesian belief function with a support function requires one to know nothing more about the support function than the *relative plausibilities of singletons*.

DEFINITION. A function $\ell:\Theta \to [0,\infty)$ is said to *express the relative plausibilities of singletons* under a support function S over Θ if

$$\ell(\theta) = c\,\mathrm{Pl}(\{\theta\}) \tag{9.3}$$

for all $\theta \in \Theta$, where Pl is the plausibility function for S and the constant c does not depend on θ.

THEOREM 9.7. *(Bayes' Theorem)** *Suppose* Bel_0 *is a Bayesian belief function over* Θ *and* S *is a support function over* Θ. *Suppose* $\ell:\Theta \to [0,\infty)$ *expresses the relative plausibilities of singletons under* S. *And suppose* $\mathrm{Bel} = \mathrm{Bel}_0 \oplus S$ *exists. Then* Bel *is Bayesian, and*

$$\mathrm{Bel}(\{\theta\}) = K\,\mathrm{Bel}_0(\{\theta\})\,\ell(\theta) \tag{9.4}$$

for all $\theta \in \Theta$, *where*

$$K = \left(\sum_{\theta \in \Theta} \mathrm{Bel}_0(\{\theta\})\,\ell(\theta)\right)^{-1}.$$

And since Bel *is Bayesian, the quantities (9.4) completely determine it;*

$$\mathrm{Bel}(A) = K \sum_{\theta \in A} \mathrm{Bel}_0(\{\theta\})\,\ell(\theta)$$

for all $A \subset \Theta$.

---

*Compare (9.4) with (1.7).

Notice that knowledge of a function ℓ satisfying (9.3) does not entail knowledge of the constant c. Hence Theorem 9.7 does indeed tell us that we need only assess the relative plausibilities of singletons under new evidence in order to combine that evidence with a Bayesian belief function. We do not need to know the values of $Pl(A)$ for A containing more than one element, and we do not even need to know the absolute values of the quantities $Pl(\{\theta\})$ for the various $\theta \in \Theta$.

Of course, when we do know a support function $S: 2^\Theta \to [0,1]$ and its plausibility function Pl, we can express the relative plausibilities of singletons by the function $\ell: \Theta \to [0,\infty)$ that is defined by $\ell(\theta) = Pl(\{\theta\})$. So (9.4) tells us in particular that

$$Bel(\{\theta\}) = K\, Bel_0(\{\theta\}) Pl(\{\theta\}),$$

where

$$K = \left(\sum_{\theta \in \Theta} Bel_0(\{\theta\}) Pl(\{\theta\})\right)^{-1}.$$

It should be noted that the relative plausibilities of singletons behave multiplicatively under combination:

> THEOREM 9.8. *Suppose* $S_1, \cdots, S_n$ *are combinable support functions, and suppose* $\ell_i : \Theta \to [0,\infty)$ *expresses the relative plausibilities of singletons under* S_i, *for* $i = 1, \cdots, n$. *Then* $\ell_1 \ell_2 \cdots \ell_n$ *expresses the relative plausibilities of singletons under* $S_1 \oplus \cdots \oplus S_n$.

So (9.4) may be used to combine any number of support functions with a Bayesian belief function Bel_0. For example, if $Pl_1, \cdots, Pl_n$ are the plausibility functions for support functions $S_1, \cdots, S_n$, and

$$Bel = Bel_0 \oplus S_1 \oplus \cdots \oplus S_n = Bel_0 \oplus (S_1 \oplus \cdots \oplus S_n)$$

exists, then Bel is Bayesian, and

$$Bel(\{\theta\}) = K\, Bel_0(\{\theta\})\, Pl_1(\{\theta\}) \cdots Pl_n(\{\theta\})$$

for all $\theta \in \Theta$, where

$$K = \left(\sum_{\theta \in \Theta} \mathrm{Bel}_0(\{\theta\})\, \mathrm{Pl}_1(\{\theta\}) \cdots \mathrm{Pl}_n(\{\theta\}) \right)^{-1} .$$

As we will see in Chapter 11, there are situations where it seems easier to assess the relative plausibilities of the singletons in a particular frame Θ than to assess degrees of support and plausibility for all the subsets of Θ. In such situations, the otherwise undesirable simplification effected by a Bayesian belief function may seem attractive, and we may even find ourselves wishing that we had such a function to combine with the evidence. After all, if there were a reasonable convention for establishing "prior" Bayesian belief functions – a convention that could be applied without making arbitrary choices that strongly affect one's final belief function – then there might be a strong case for the use of such "priors." But as we learned in §7 of Chapter 1, there is no such convention.

The basic difficulty is that the only natural convention for establishing a prior Bayesian belief function is strongly dependent on the frame of discernment and is sensitive to refinement or coarsening. If we begin with a particular frame Θ that has n elements, then the only symmetrical Bayesian belief function in view is the *uniform* one over Θ – the one that assigns degree of belief $\frac{1}{n}$ to each singleton of Θ. This function is directly symmetrical, and it is symmetrical with respect to the infinite weights of evidence, for it corresponds to the focusing of equal infinite weights on each singleton. (See Case 1 of Example 9.1.) But it depends crucially on the frame of discernment Θ that one has chosen; the same convention for a compatible frame Ω may yield a prior over Ω that is radically incompatible with the one over Θ. And the radical differences among different uniform priors persist in the differences among the "posterior" belief functions that they yield when combined with the evidence. Being surrogates for imaginary infinite weights of evidence, such priors can overwhelm any actual finite weights of evidence to yield almost any degrees of belief one desires.

EXAMPLE 9.2. *Incompatible Uniform Priors.* Consider the frame of discernment

$$\Theta = \{LI, LO, RI, RO\}$$

of Example 4.1, and suppose that the actual evidence produces a simple support function S focused on $\{LI, LO\}$ with $S(\{LI, LO\}) = \frac{1}{2}$. Then the relative plausibilities of singletons under S are expressed by the function $\ell : \Theta \to [0, \infty)$, where $\ell(LI) = \ell(LO) = 1$ and $\ell(RI) = \ell(RO) = \frac{1}{2}$.

If we let $Bel_0 : 2^\Theta \to [0, 1]$ denote the uniform Bayesian prior over Θ, then $Bel_0(\{\theta\}) = \frac{1}{4}$ for all $\theta \in \Theta$, and the Bayesian belief function $Bel = Bel_0 \oplus S$ is given by $Bel(\{LI\}) = Bel(\{LO\}) = \frac{1}{3}$ and $Bel(\{RI\}) = Bel(\{RO\}) = \frac{1}{6}$. Notice that $Bel(\{LI, LO\}) = \frac{2}{3}$ and $Bel(\{RI, RO\}) = \frac{1}{3}$; these might be excused as being not too far from the true values $S(\{LI, LO\}) = \frac{1}{2}$ and $S(\{RI, RO\}) = 0$.

But now let us consider a refinement $\omega : 2^\Theta \to 2^\Omega$ such that $\omega(\{LI\})$ and $\omega(\{LO\})$ are singletons but $\omega(\{RI\})$ and $\omega(\{RO\})$ each have 98 elements. And for simplicity, let us assume that Θ exhausts the impact of the evidence on Ω, so that the support function $S^\Omega : 2^\Omega \to [0, 1]$ is the vacuous extension of S to Ω. Then the relative plausibilities of singletons under S^Ω are expressed by the function $\ell^\Omega : \Omega \to [0, \infty)$, where $\ell^\Omega(\zeta) = 1$ if $\zeta \in \omega(\{LI, LO\})$, and $\ell^\Omega(\zeta) = \frac{1}{2}$ if $\zeta \in \omega(\{RI, RO\})$.

If we let $Bel_0^\Omega : 2^\Omega \to [0, 1]$ denote the uniform Bayesian prior over Ω, then $Bel_0^\Omega(\{\zeta\}) = \frac{1}{198}$ for all $\zeta \in \Omega$, and the Bayesian belief function $Bel^\Omega = Bel_0^\Omega \oplus S^\Omega$ is given by $Bel^\Omega(\{\zeta\}) = \frac{1}{100}$ for all $\zeta \in \omega(\{LI, LO\})$ and $Bel^\Omega(\{\zeta\}) = \frac{1}{200}$ for all $\zeta \in \omega(\{RI, RO\})$. In particular, the degree of belief for $\{LI, LO\}$ is now .02, while the degree of belief for $\{RI, RO\}$ is now .98. ∎

§6. Mathematical Appendix

Proof of Theorem 9.1. Suppose, indeed, that the function Bel on 2^Θ is the limit of a sequence $\mathrm{Bel}_1, \mathrm{Bel}_2, \cdots$ of belief functions over Θ. Then Bel certainly maps 2^Θ into $[0,1]$, and we can establish that Bel is a belief function by establishing that it obeys the three rules of Theorem 1:

(1) $\mathrm{Bel}(\emptyset) = \lim_{i\to\infty} \mathrm{Bel}_i(\emptyset) = \lim_{i\to\infty} 0 = 0.$

(2) $\mathrm{Bel}(\Theta) = \lim_{i\to\infty} \mathrm{Bel}_i(\Theta) = \lim_{i\to\infty} 1 = 1.$

(3) For every positive integer n and every collection $A_1, \cdots, A_n$ of subsets of Θ,

$$\begin{aligned}
\mathrm{Bel}(A_1 \cup \cdots \cup A_n) &= \lim_{i\to\infty} \mathrm{Bel}_i(A_1 \cup \cdots \cup A_n) \\
&\geq \lim_{i\to\infty} \sum_{\substack{I \subset \{1,\cdots,n\} \\ I \neq \emptyset}} (-1)^{|I|+1} \mathrm{Bel}_i\Big(\bigcap_{j\in I} A_j\Big) \\
&= \sum_{\substack{I \subset \{1,\cdots,n\} \\ I \neq \emptyset}} (-1)^{|I|+1} \lim_{i\to\infty} \mathrm{Bel}_i\Big(\bigcap_{j\in I} A_j\Big) \\
&= \sum_{\substack{I \subset \{1,\cdots,n\} \\ I \neq \emptyset}} (-1)^{|I|+1} \mathrm{Bel}\Big(\bigcap_{j\in I} A_j\Big) . \blacksquare
\end{aligned}$$

Proof of Theorem 9.2. Let m denote the basic probability assignment for Bel. Since Bel is not a support function, $m(\Theta) = 0$.

For every positive integer k, define a basic probability assignment $m_k : 2^\Theta \to [0,1]$ by

$$m_k(A) = \frac{k}{k+1}\, m(A)$$

for all proper subsets A of Θ, and

$$m_k(\Theta) = \frac{1}{k+1} .$$

Let Bel_k denote the belief function for m_k. Since Bel_k gives its core Θ a positive basic probability number, it is a support function.

Now all the Bel_k will have the same focal elements – say Θ and $B_1, \cdots, B_n$. According to Lemma 7.4, there exists a refining Ω of Θ and there exist disjoint subsets $A_1, \cdots, A_n$ of Ω such that $\bar{\theta}(A_i) = B_i$ for $i = 1, \cdots, n$.

For each k, let $m'_k : 2^\Omega \to [0,1]$ be a basic probability assignment with focal elements $A_1, \cdots, A_n$, and Ω; set $m'_k(A_i) = m_k(B_i)$ for $i = 1, \cdots, n$, and $m'_k(\Omega) = m_k(\Theta)$. Let S_k denote the belief function for m_k. By the proof of Theorem 7.1, S_k is a separable support function, and $Bel_k = S_k | 2^\Theta$.

As k tends to infinity, m_k converges to m, and m'_k tends to the basic probability assignment $m' : 2^\Omega \to [0,1]$ given by $m'(A_i) = m(B_i)$ for $i = 1, \cdots, n$, and $m'(A) = 0$ for all other A. Hence the Bel_k tend to Bel, the S_k also tend to a limit, and

$$Bel = \lim_{k \to \infty} Bel_k = \lim_{k \to \infty} (S_k | 2^\Theta) = \left(\lim_{k \to \infty} S_k \right) | 2^\Theta . \blacksquare$$

LEMMA 9.1. *Suppose* Bel *is the limit of a sequence* $Bel_1, Bel_2, \cdots$ *of belief functions, and denote the corresponding commonality functions by* $Q, Q_1, Q_2, \cdots$. *Then the sequence* $Q_1, Q_2, \cdots$ *converges to* Q.

Proof of Lemma 9.1. By Theorem 2.4,

$$\begin{aligned} Q(A) &= \sum_{B \subset A} (-1)^{|B|} Bel(\bar{B}) \\ &= \sum_{B \subset A} (-1)^{|B|} \lim_{i \to \infty} Bel_i(\bar{B}) \\ &= \lim_{i \to \infty} \sum_{B \subset A} (-1)^{|B|} Bel_i(\bar{B}) \\ &= \lim_{i \to \infty} Q_i(A) \end{aligned}$$

for all $A \subset \Theta$. $\blacksquare$

Proof of Theorem 9.3. For each i, $i=1,\cdots,n$, let Q_i denote S_i's commonality function, and let w_i denote its assessment of evidence.

Let Bel denote the limit of the sequence $S_1, S_2, \cdots$. By Theorem 9.1, Bel is a belief function. Let Q denote its commonality function, and let $\mathcal{C}$ denote its core.

Let us establish the theorem by proving its contrapositive. In other words, let us assume that there exists $\theta_0 \in \Theta$ for which $V_i(\{\theta_0\})$ does not tend to infinity, and then prove that Bel is a separable support function.

If $V_i(\{\theta_0\})$ does not tend to infinity with i, then there is a subsequence $V_{i_1}(\{\theta_0\}), V_{i_2}(\{\theta_0\}), \cdots$ that is bounded from above. Since this subsequence will also correspond to a sequence of separable support functions tending to Bel, we may assume that all the $V_i(\{\theta_0\})$ are bounded – i.e., that there exists a constant c such that

$$V_i(\{\theta_0\}) \leq c$$

for all i.

By (5.10),

$$Q_i(\{\theta_0\}) = K_i e^{-v_i(\{\theta_0\})} ,$$

where the constants K_i are all greater than or equal to one. So

$$Q_i(\{\theta_0\}) \geq e^{-v_i(\{\theta_0\})} \geq e^{-c} > 0$$

for all i. By Lemma 9.1, $Q(\{\theta_0\})$ is the limit of the $Q_i(\{\theta_0\})$; so it is positive as well. It follows that for each $\theta \in \Theta$,

$$\frac{Q_i(\{\theta\})}{Q_i(\{\theta_0\})} \longrightarrow \frac{Q(\{\theta\})}{Q(\{\theta_0\})}$$

as i tends to infinity. But $Q(\{\theta\}) > 0$ if and only if $\theta \in \mathcal{C}$, and

$$\frac{Q_i(\{\theta\})}{Q_i(\{\theta_0\})} = e^{v_i(\{\theta_0\}) - v_i(\{\theta\})}$$

for all $\theta \in \Theta$. Hence $V_i(\{\theta\})$ tends to infinity as i tends to infinity if and only if $\theta \in \overline{\mathcal{C}}$. And $V_i(\{\theta\})$ is bounded for every $\theta \in \mathcal{C}$. Now

$$V_i(\mathcal{C}) = \sum_{\substack{B \subset \Theta \\ \mathcal{C} \not\subset B}} w_i(B) \leq \sum_{\theta \in \mathcal{C}} \sum_{\substack{B \subset \Theta \\ \theta \notin B}} w_i(B) = \sum_{\theta \in \mathcal{C}} V_i(\{\theta\}) .$$

So $V_i(\mathcal{C})$ is bounded as i tends to infinity. And since

$$Q_i(\mathcal{C}) = K_i e^{-V_i(\mathcal{C})} ,$$

it follows that $Q(\mathcal{C})$, the limit of the quantities $Q_i(\mathcal{C})$, is positive. So $Q(B) > 0$ for all $B \subset \mathcal{C}$.

Define a function $w : 2^\Theta \to [0, \infty]$ by

$$w(A) = \sum_{\substack{B \subset \mathcal{C} \\ A \subset B}} (-1)^{|B-A|} \log Q(B)$$

$$= \lim_{i \to \infty} \sum_{\substack{B \subset \mathcal{C} \\ A \subset B}} (-1)^{|B-A|} \log Q_i(B)$$

$$= \lim_{i \to \infty} w_i(A) \geq 0$$

for all proper non-empty $A \subset \mathcal{C}$, $w(\mathcal{C}) = \infty$ and $w(A) = 0$ for all $A \not\subset \mathcal{C}$. Notice that w qualifies as an assessment of evidence. And by Lemma 5.2, Q is the commonality function determined by w. So Bel is a separable support function. ∎

Proof of Theorem 9.4. Since $\text{Bel}(A) > 0$ and $\text{Bel}(\bar{A}) > 0$, both A and $\bar{A}$ must contain focal elements of Bel. Suppose $B_1 \subset A$ and $B_2 \subset \bar{A}$ are focal elements and choose $\theta_1 \in B_1$ and $\theta_2 \in B_2$. Then

$$Q(\{\theta_1\}) \geq m(B_1) > 0$$

and

$$Q(\{\theta_2\}) \geq m(B_2) > 0 .$$

So θ_1 and θ_2 are both in the core $\mathcal{C}$ of Bel. And $\{\theta_1, \theta_2\} \subset \mathcal{C}$.

Since $\mathrm{Bel}(A) + \mathrm{Bel}(\bar{A}) = 1$, every focal element of Bel must be contained in A or in $\bar{A}$. Hence no focal element of Bel can contain $\{\theta_1, \theta_2\}$; $Q(\{\theta_1, \theta_2\}) = 0$. Hence $Q(\mathcal{C}) = 0$. So by Theorem 7.1, Bel is not a support function, whence, by Theorem 9.2, Bel is a quasi support function. ∎

Proof of Theorem 9.5. By Theorem 2.8, a Bayesian belief function awards zero commonality numbers to all non-singletons. By Theorem 7.1, a belief function is a support function if and only if it awards a positive commonality number to its core. So a Bayesian belief function is a support function if and only if its core is a singleton. ∎

LEMMA 9.2. *Suppose* $\mathrm{Bel}: 2^{\Theta} \to [0,1]$ *is a belief function. Denote its core by* $\mathcal{C}$ *and its commonality function by* Q. *Then* Bel *is Bayesian if and only if* $Q(\{\theta, \theta'\}) = 0$ *whenever* $\theta \in \mathcal{C}$, $\theta' \in \Theta$, *and* $\theta \neq \theta'$.

Suppose $\bar{\theta}: 2^{\Theta} \to 2^{\Theta_0}$ *is an outer reduction. Then* $\mathrm{Bel}|2^{\Theta_0}$ *is Bayesian if and only if* $Q(\{\theta,\theta'\})=0$ *whenever* $\theta \in \mathcal{C}$, $\theta' \in \Theta$, *and* $\bar{\theta}(\{\theta\}) \neq \bar{\theta}(\{\theta'\})$.

Proof of Lemma 9.2. According to Theorem 2.8, Bel is Bayesian if and only if $Q(A) = 0$ for all $A \subset \Theta$ that are not singletons. Since $Q(A) \leq Q(B)$ whenever $B \subset A$, it follows that Bel is Bayesian if and only if $Q(\{\theta,\theta'\}) = 0$ for all doubletons $\{\theta,\theta'\} \subset \Theta$. Since Q(A) is necessarily zero whenever A includes a point not in the core, it follows that Bel is Bayesian if and only if $Q(\{\theta,\theta'\}) = 0$ whenever $\theta \in \mathcal{C}$, $\theta' \in \mathcal{C}$, and $\theta \neq \theta'$.

Let us now turn to the second paragraph of the lemma. Let m denote the basic probability assignment for Bel, let Q_0 denote the commonality function for $\mathrm{Bel}|2^{\Theta_0}$, and let $\omega: 2^{\Theta_0} \to 2^{\Theta}$ denote the refining. Then by the argument of the preceding paragraph, $\mathrm{Bel}|2^{\Theta}$ is Bayesian if and only if $Q_0(\{\theta_0, \theta'_0\}) = 0$ for all doubletons $\{\theta_0, \theta'_0\} \subset \Theta_0$.

By Theorem 6.9,

$$Q_0(\{\theta_0, \theta'_0\}) = \sum_{\substack{B \subset \Theta \\ \{\theta_0, \theta'_0\} \subset \bar{\theta}(B)}} m(B) .$$

And $\{\theta_0, \theta'_0\} \subset \bar{\theta}(B)$ if and only if $\omega(\{\theta_0\})$ and $\omega(\{\theta'_0\})$ both have non-empty intersections with B. So $Q_0(\{\theta_0, \theta'_0\}) = 0$ for all doubletons $\{\theta_0, \theta'_0\} \subset \Theta_0$ if and only if $m(B) = 0$ for all $B \subset \Theta$ that intersect more than one $\omega(\{\theta_0\})$ – i.e., if and only if $m(B) = 0$ for all B containing elements θ and θ' such that $\bar{\theta}(\{\theta\}) \neq \bar{\theta}(\{\theta'\})$. But m(B) may be replaced by Q(B) in this latter condition without changing its meaning. So $Bel|2^{\Theta}$ is Bayesian if and only if $Q(\{\theta, \theta'\}) = 0$ whenever $\bar{\theta}(\{\theta\}) \neq \bar{\theta}(\{\theta'\})$. Since $Q(\{\theta, \theta'\})$ is necessarily zero if $\theta \notin \mathcal{C}$, the second paragraph of the lemma follows. ∎

Proof of Theorem 9.6.

(i) Let Q denote the commonality function for Bel, and let Q_i denote the commonality function for the separable support function determined by w_i. According to Lemma 9.1, Q_i converges to Q. This means in particular that

$$\frac{Q_i(\{\theta, \theta'\})}{Q_i(\{\theta\})} \text{ converges to } \frac{Q(\{\theta, \theta'\})}{Q(\{\theta\})}$$

whenever $\theta \in \mathcal{C}$ and $\theta' \in \Theta$. But we know from (5.5) that

$$\frac{Q_i(\{\theta, \theta'\})}{Q_i(\{\theta\})} = \exp\left(-\sum_{\substack{B \subset \Theta \\ \theta \in B \\ \theta' \notin B}} w_i(B)\right)$$

whenever $\theta \in \mathcal{C}$ and i is sufficiently large to ensure that $Q_i(\{\theta\}) > 0$. Applying Lemma 9.2, we may conclude that Bel is Bayesian if and only if

$$\lim_{i \to \infty} \sum_{\substack{B \subset \Theta \\ \theta \in B \\ \theta' \notin B}} w_i(B) = \infty \tag{9.5}$$

whenever $\theta \in \mathcal{C}$, $\theta' \in \Theta$, and $\theta \neq \theta'$. So our task in proving the first part of the theorem reduces to showing that (9.1) holds for all $\theta \in \mathcal{C}$ if and only if (9.5) holds whenever $\theta \in \mathcal{C}$, $\theta' \in \Theta$, and $\theta \neq \theta'$.

Suppose (9.5) holds whenever $\theta \in \mathcal{C}$, $\theta' \in \Theta$, and $\theta \neq \theta'$. Fixing $\theta \in \mathcal{C}$, let $\theta_1, \cdots, \theta_n$ be all the other elements of Θ. Then for any constant K, there exists N sufficiently large that

$$\sum_{\substack{B \subset \Theta \\ \theta \in B \\ \theta_j \notin B}} w_i(B) \geq K|2^{\Theta}| \tag{9.6}$$

for all $i \geq N$ and all j, $j = 1, \cdots, n$. So for each i, $i \geq N$, we can choose $B_{i1}, \cdots, B_{in}$ such that $\theta \in B_{ij}$, $\theta_j \notin B_{ij}$, and

$$w_i(B_{ij}) \geq K$$

for $j = 1, \cdots, n$. If we set $\mathcal{Q}_i = \{B_{i1}, \cdots, B_{in}\}$, then $\cap \mathcal{Q}_i = \{\theta\}$, and

$$\min_{B \in \mathcal{Q}_i} w_i(B) \geq K$$

for all $i \geq N$. Since the element $\theta \in \mathcal{C}$ and the constant K are arbitrary, (9.1) follows.

Now suppose (9.1) holds for all $\theta \in \mathcal{C}$, and fix $\theta \in \mathcal{C}$. Then for every constant K there exists an integer N such that for every $i \geq N$ there exists $\mathcal{Q}_i \subset 2^{\Theta}$ such that $\cap \mathcal{Q}_i = \{\theta\}$ and

$$\min_{B \in \mathcal{Q}_i} w_i(B) \geq K .$$

Suppose $\theta' \epsilon \Theta$ and $\theta' \neq \theta$. Then there must exist $B_i \epsilon \mathcal{A}_i$ such that $\theta' \notin B_i$. So

$$\sum_{\substack{B \subset \Theta \\ \theta \epsilon B \\ \theta' \notin B}} w_i(B) \geq w_i(B_i) \geq K \tag{9.7}$$

for $i \geq N$. Since K is arbitrary, (9.5) follows.

(ii) Now let us turn to the second part of the theorem. Using Lemma 9.2 and arguing as in the first paragraph of this proof, we find that $Bel|2^{\Theta}$ is Bayesian if and only if (9.5) holds whenever $\theta \epsilon \mathcal{C}$, $\theta' \epsilon \Theta$, and $\overline{\theta}(\{\theta\}) \neq \overline{\theta}(\{\theta'\})$. By Theorem 6.9,

$$\overline{\theta}(\mathcal{C}) = \bigcup_{\theta \epsilon \mathcal{C}} \overline{\theta}(\{\theta\})$$

is the core of $Bel|2^{\Theta_0}$. So θ_0 is in the core of $Bel|2^{\Theta}$ if and only if $\theta_0 = \overline{\theta}(\{\theta\})$ for some $\theta \epsilon \mathcal{C}$. And our task reduces to showing that

$$\lim_{i \to \infty} \max_{\substack{\mathcal{A} \subset 2^{\Theta} \\ \emptyset \neq \cap\mathcal{A} \subset \omega(\overline{\theta}(\{\theta\}))}} \min_{B \epsilon \mathcal{A}} w_i(B) = \infty \tag{9.8}$$

for all $\theta \epsilon \mathcal{C}$ if and only if (9.5) holds whenever $\theta \epsilon \mathcal{C}$, $\theta' \epsilon \Theta$, and $\overline{\theta}(\{\theta\}) \neq \overline{\theta}(\{\theta'\})$.

Suppose (9.5) holds whenever $\theta \epsilon \mathcal{C}$, $\theta' \epsilon \Theta$, and $\overline{\theta}(\{\theta\}) \neq \overline{\theta}(\{\theta'\})$. Fixing $\theta \epsilon \mathcal{C}$, let $\theta_1, \cdots, \theta_n$ be all the elements of Θ such that $\overline{\theta}(\{\theta\}) \neq \theta(\{\theta_i\})$. Then for any constant K, there exists N sufficiently large that (9.6) holds for all $i \geq N$ and all j, $j = 1, \cdots, n$. So for each i, $i \geq N$, we can choose $B_{i1}, \cdots, B_{in}$ such that $\theta \epsilon B_{ij}$, $\theta_j \notin B_{ij}$, and

$$w_i(B_{ij}) \geq K$$

for $j = 1, \cdots, n$. If we set $\mathcal{A}_i = \{B_{i1}, \cdots, B_{in}\}$, then $\cap\mathcal{A}_i \neq \emptyset$, for $\{\theta\} \subset \mathcal{A}_i$. And

$$\min_{B \epsilon \mathcal{A}_i} w_i(B) \geq K$$

for all $i \geq N$. Furthermore, $\theta' \notin \cap \mathcal{A}_i$ if $\overline{\theta}(\{\theta\}) \neq \overline{\theta}(\{\theta'\})$. But $\overline{\theta}(\{\theta\}) = \overline{\theta}(\{\theta'\})$ only if $\theta' \in \omega(\overline{\theta}(\{\theta\}))$. So $\cap \mathcal{A}_i \subset \omega(\overline{\theta}(\{\theta\}))$. Since the element $\theta \in \mathcal{C}$ and the constant K are arbitrary, (9.8) follows.

Now suppose (9.8) holds for all $\theta \in \mathcal{C}$, and fix $\theta \in \mathcal{C}$. Then for every constant K there exists an integer N such that for every $i \geq N$ there exists $\mathcal{A}_i \subset 2^\Theta$ such that $\emptyset \neq \cap \mathcal{A}_i \subset \omega(\overline{\theta}(\{\theta\}))$ and

$$\min_{B \in \mathcal{A}_i} w_i(B) \geq K .$$

Suppose $\theta' \in \Theta$ and $\overline{\theta}(\{\theta'\}) \neq \overline{\theta}(\{\theta\})$. Since $\overline{\theta}(\{\theta'\})$ and $\overline{\theta}(\{\theta\})$ are singletons, they must be disjoint, and since $\theta \in \omega(\overline{\theta}(\{\theta\}))$, it follows that $\theta' \notin \omega(\overline{\theta}(\{\theta\}))$. Hence $\theta' \notin \cap \mathcal{A}_i$, and there must exist $B_i \in \mathcal{A}_i$ such that $\theta' \notin B_i$. So (9.7) holds. Since K is arbitrary, (9.5) follows. ∎

Proof of Theorem 9.7. Let Q, Q_0 and Q' denote the commonality functions for S, Bel_0 and Bel, respectively. Let P^* and P_0^* denote the upper probability functions for Bel and Bel_0, and let Pl denote the plausibility function for S.

Bel is Bayesian by Theorem 3.7. And by Theorems 2.8 and 3.3,

$$\begin{aligned} Bel(\{\theta\}) &= P^*(\{\theta\}) = Q'(\{\theta\}) \\ &= K_0 Q_0(\{\theta\}) Q(\{\theta\}) \\ &= K_0 P_0^*(\{\theta\}) Pl(\{\theta\}) \\ &= K_0 Bel_0(\{\theta\}) c\, \ell(\theta) \\ &= K\, Bel_0(\{\theta\}) \ell(\theta) \end{aligned}$$

for all $\theta \in \Theta$, where the constant $K = K_0 c$ does not depend on θ. The formula for Bel(A) follows by Lemma 2.6, and the formula for K follows because

$$1 = Bel(\Theta) = \sum_{\theta \in \Theta} Bel(\{\theta\}) = K \sum_{\theta \in \Theta} Bel_0(\{\theta\}) \ell(\theta) .\ ∎$$

Proof of Theorem 9.8. Let $Q_1,\cdots,Q_n$ and Q denote the commonality functions for $S_1,\cdots,S_n$ and $S_1\oplus\cdots\oplus S_n$. And let $Pl_1,\cdots,Pl_n$ and Pl denote the plausibility functions. Then $Pl(\{\theta\}) = Q(\{\theta\})$ for all $\theta \epsilon \Theta$, and $Pl_i(\{\theta\}) = Q_i(\{\theta\})$ for all $\theta \epsilon \Theta$ and all i. And there exist constants $c_1,\cdots,c_n$ such that $\ell_i(\theta) = c_i Pl_i(\{\theta\})$ for all $\theta \epsilon \Theta$ and all i. So by (3.5),

$$\begin{aligned}\ell_1(\theta)\cdots\ell_n(\theta) &= (c_1 Pl_1(\{\theta\}))\cdots(c_n Pl_n(\{\theta\}))\\ &= (c_1\cdots c_n)Q_1(\{\theta\})\cdots Q_n(\{\theta\})\\ &= (c_1\cdots c_n)\frac{1}{K}\,Q(\{\theta\})\\ &= (c_1\cdots c_n)\frac{1}{K}\,Pl(\{\theta\})\end{aligned}$$

for all $\theta \epsilon \Theta$. ∎

CHAPTER 10. CONSONANCE

> It often happens in science that while data are scarce, interpretation seems easy, but as the number of data grows, consistent argument becomes more and more difficult.
>
> HITOSHI TAKEUCHI, SEIYA UYEDA, and HIROO KANAMORI.*

At the opposite extreme from quasi support functions we find the consonant support functions. These functions are true support functions; in fact, they are separable support functions. And they are distinguished by their failure to betray even a hint of conflict in the evidence. The evidence underlying a consonant support function is not necessarily homogeneous in the sense that the evidence underlying a simple support function is homogeneous. But it can be described as "pointing in a single direction"; it is heterogeneous only in that it varies in the precision of its focus.

In the first two sections of this chapter, we study the mathematics of consonant support functions. In §3 we discuss work by two scholars, L. Jonathan Cohen and G. L. S. Shackle, who seem to have regarded all evidence as consonant. And in §4 we study a type of evidence that it does seem natural to regard as consonant: inferential evidence.

§1. Consonant Support Functions

A belief function is said to be *consonant* if its focal elements are nested – i.e., if its focal elements can be arranged in order so that each

---

*The quotation is from p. 180 of *Debate about the Earth: Approach to Geophysics through Analysis of Continental Drift* (Revised Edition), published in English in 1970.

is contained in the following one. It is easy to construct examples of consonant belief functions. Every simple support function is one, and a slightly more complicated one is exhibited in Example 4.2.

The following theorem provides some insight into the special nature of consonant belief functions and shows in particular that they are always separable support functions.

THEOREM 10.1. *Suppose* $\mathrm{Bel}: 2^{\Theta} \to [0,1]$ *is a belief function, with upper probability function* P^* *and commonality function* Q. *Then the following assertions are all equivalent:*

(1) Bel *is consonant.*

(2) $\mathrm{Bel}(A \cap B) = \min(\mathrm{Bel}(A), \mathrm{Bel}(B))$ *for all* $A, B \subset \Theta$.

(3) $P^*(A \cup B) = \max(P^*(A), P^*(B))$ *for all* $A, B \subset \Theta$.

(4) $P^*(A) = \max\limits_{\theta \in A} P^*(\{\theta\})$ *for all non-empty* $A \subset \Theta$.

(5) $Q(A) = \min\limits_{\theta \in A} Q(\{\theta\})$ *for all non-empty* $A \subset \Theta$.

(6) *There exist a positive integer* n *and simple support functions* $S_1, \cdots, S_n$ *such that* $\mathrm{Bel} = S_1 \oplus \cdots \oplus S_n$ *and the focus of* S_i *is contained in the focus of* S_j *whenever* $i < j$.

Assertion (6) of this theorem reveals the non-conflicting and consonant character of the evidence underlying a consonant support function. All the evidence points in the same direction, and it is internally heterogeneous only to the extent that it varies in its precision. This understanding needs to be qualified only slightly: the evidence need not have a completely nested character, provided that its non-nested aspects are obscured by infinite weights of evidence.

THEOREM 10.2. *Suppose* $S_1, \cdots, S_n$ *are non-vacuous simple support functions with foci* $A_1, \cdots, A_n$, *respectively, and suppose* $S = S_1 \oplus \cdots \oplus S_n$ *is consonant. Let* $\mathcal{C}$ *denote the core of* S. *Then the sets* $A_i \cap \mathcal{C}$, $i = 1, \cdots, n$, *are nested; they can be ordered so that each is contained in the following one.*

Suppose Bel is a consonant support function over Θ. Then by (2) of Theorem 10.1,

$$0 = \mathrm{Bel}(\emptyset) = \mathrm{Bel}(A \cap \bar{A}) = \min(\mathrm{Bel}(A), \mathrm{Bel}(\bar{A}))$$

– i.e., either $\mathrm{Bel}(A) = 0$ or $\mathrm{Bel}(\bar{A}) = 0$ – for every $A \subset \Theta$; Bel never accords positive degrees of belief to both sides of a dichotomy. This property is shared by some non-consonant or *dissonant* belief functions, but only consonant support functions retain the property under all conditionings.

THEOREM 10.3. *A belief function* Bel *over a frame* Θ *is consonant if and only if*

$$\min(\mathrm{Bel}(A|B), \mathrm{Bel}(\bar{A}|B)) = 0$$

for all $A \subset \Theta$ *and all* $B \subset \Theta$ *such that* $\mathrm{Bel}(\bar{B}) < 1$. *(Recall from Theorem 3.6 that* $\mathrm{Bel}(\cdot|B)$ *is the orthogonal sum of* Bel *and the simple support function focusing unit support on* B.)

It was with this theorem in mind that I asserted in the introduction above that a consonant support function fails to betray even a hint of conflict in the evidence.

§2. The Contour Function

Though the consonant support functions contrast sharply with the Bayesian belief functions in being based on non-conflicting evidence, they resemble the Bayesian belief functions in one important practical respect: they are completely determined by the plausibilities they award to singletons.

This means that from a practical point of view a consonant support function is essentially a point function rather than a set function. In fact, it is most natural to express a consonant support function over Θ by its *contour function* $f: \Theta \to [0,1]$, obtained from the plausibility function Pl by

$$f(\theta) = \mathrm{Pl}(\{\theta\}) .$$

From this function one may immediately recover both Pl and the commonality function Q, for since Q always agrees with Pl on singletons, Theorem 10.1 tells us that

$$\mathrm{Pl}(A) = \max_{\theta \in A} f(\theta)$$

and

$$Q(A) = \min_{\theta \in A} f(\theta)$$

for all non-empty $A \subset \Theta$.

As its name suggests, the contour function may be given a geometric interpretation. Figures 10.1 and 10.2 exhibit this interpretation in the case of a consonant belief function with five nested focal elements $A_1 \subset A_2 \subset A_3 \subset A_4 \subset A_5$, the first being a singleton and each having the basic probability number $\frac{1}{5}$. The picture of these nested focal elements reminds one of a hill, with the singleton A_1 at its peak and the boundaries of the other focal elements forming contours around that peak. The values of the contour function, shown in Figure 10.2, correspond in this image to the height of the points. (Notice, though, that our hill is quite sharply terraced; it is level between each pair of contour lines, jumping .2 at each contour line. In particular, the singleton A_1 stands .2 above the rest of A_2!)

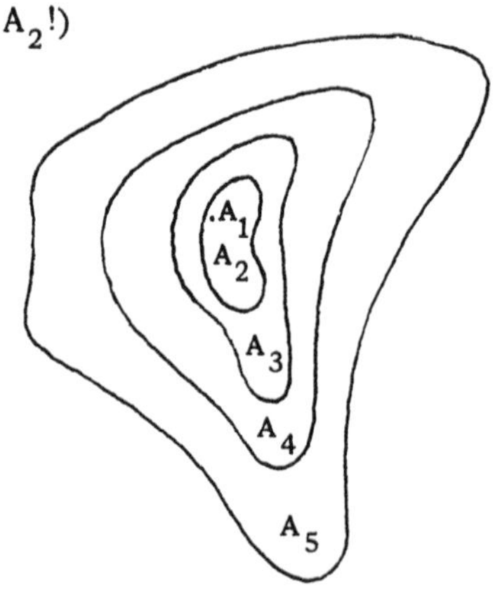

Figure 10.1

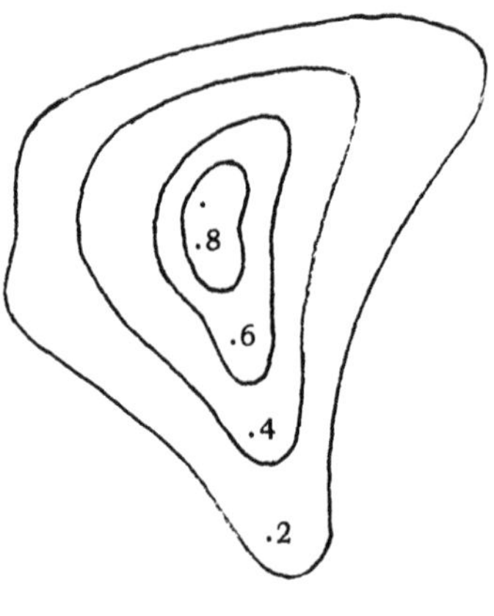

Figure 10.2

This geometric interpretation of the contour function can be helpful in picturing the movement of our mobile probability masses. Suppose, indeed, that our probability mass is originally distributed along a vertical pole of unit length, standing under the peak A_1. Then if we require each increment of this probability mass to stay at its original level and also to stay under the hill, then we obtain an accurate representation of how the probability masses are permitted to move over Θ.*

§3. The Embarrassment of Dissonance

Consonant support functions are attractive for their simplicity, and one might be tempted to argue that degrees of support based on evidence *ought* to be consonant – that dissonance is always a symptom of some mistake in assessing the evidence, and that such mistakes call for immediate correction. After all, if two items of evidence contradict each other, must not one of them be wrong?

In fact, at least two scholars have defended rules for degrees of belief that come down to our rules for consonant support functions. These scholars, the philosopher L. Jonathan Cohen and the economist G. L. S. Shackle, have not had the whole theory of the present essay in view, but they seem to have thought of evidence and belief as consonant, and they have pointed out the contrast between their rules and Bayes' rules.**

---

*In the example of Figure 10.1, each focal element is "connected," so that a probability mass can move between any two points in its focal element by a continuous motion under the hill. This should not, however, be understood as an essential element of the picture. If a focal element is disconnected, then its probability mass is to be allowed to jump from one part of it to another.

This geometric picture can only sharpen the reader's desire to generalize our theory so that infinite frames of discernment and continuous contour functions are permitted. In fact much of the theory of this chapter generalizes to the infinite case quite easily. But such a generalization of the whole of our theory will require considerable labor and must remain outside the bounds of this essay.

**Mr. Cohen has presented his ideas at length in *The Implications of Induction* and more briefly and lucidly in "A Note on Inductive Logic" in Volume LXX of *The Journal of Philosophy*. Mr. Shackle's ideas are most accessible in his *Decision, Order and Time in Human Affairs*, especially Chapters 9, 10, and 11. I am grateful to the friends who brought the work of these two scholars to my attention – Paul Benacerraf in the case of Mr. Cohen and Burton Singer in the case of Mr. Shackle.

There are several ways that one can identify the class of consonant support functions without being cognizant of the general theory of belief functions. Witness these two theorems:

THEOREM 10.4. *Suppose* Θ *is a finite set. Then a function* $S: 2^\Theta \to [0,1]$ *is a consonant support function if and only if* S *obeys these three rules*:

(1) $S(\emptyset) = 0$.

(2) $S(\Theta) = 1$.

(3) $S(A \cap B) = \min(S(A), S(B))$ *for all* $A, B \subset \Theta$.

THEOREM 10.5. *Suppose* Θ *is a finite set. Then a function* $\mathrm{Bel}: 2^\Theta \to [0,1]$ *is a consonant belief function if and only if* $\mathrm{Bel}(\Theta) = 1$ *and the function* $\mathrm{Dou}: 2^\Theta \to [0,1]$ *given by* $\mathrm{Dou}(A) = \mathrm{Bel}(\bar{A})$ *obeys these two rules*:

(1) $\mathrm{Dou}(A) = \min_{\theta \in A} \mathrm{Dou}(\{\theta\})$ *for all non-empty* $A \subset \Theta$.

(2) *There exists* $\theta \in \Theta$ *with* $\mathrm{Dou}(\{\theta\}) = 0$.

If one studies functions obeying rules (1), (2), and (3) of Theorem 10.4, then one is studying consonant support functions — whether or not one is aware of Dempster's rule of combination and the other elements of our mathematical theory of evidence. And since rules (1) and (2) are essentially conventions, rule (3) is sufficient by itself to bring the object of one's study well into the realm of our theory. Similarly, any study of functions satisfying rules (1) and (2) of Theorem 10.5 will produce echoes of our general theory.

The picture of evidence painted in Mr. Cohen's writings seems to be a picture of consonant evidence, and thus it is not surprising that he advocates rule (3) of Theorem 10.4 explicitly and emphatically. This is not to say that he explicitly considers what dissonant evidence might be like and rejects the possibility of such evidence. But at every stage of an investigation he expects to see the evidence pointing consistently to some possibility or hypothesis that has avoided any impugnation by the evidence. I do not fully understand his reasons for this expectation, but there are hints that he would have us react to apparent dissonance by changing our frame Θ so as to introduce new possibilities that escape impugnation: "Where the outcome fails to accord exactly with the initial hypothesis, an appropriately modified hypothesis is substituted so as to maintain the growing immunity to reasonable doubt."*

Mr. Shackle seldom writes of evidence; he relies instead on direct speculation about how partial beliefs ought to behave. He usually writes about the "degree of potential surprise" associated with a proposition, a quantity that corresponds to the degree of doubt $\mathrm{Dou}(A) = \mathrm{Bel}(\bar{A})$ of our theory, both in its intuitive meaning and its relation to the degree of belief. And he explicitly imposes on these degrees of potential surprise the two rules listed in Theorem 10.5. He is aware that these rules exclude assigning positive degrees of potential surprise (or positive degrees of belief) to both sides of a dichotomy, and he argues that this is appropriate: "To assign greater than zero degrees of potential surprise to both the hypothesis and its contradictory would ··· betray an unresolved mental confusion."**

It is easy to share the desire of these scholars to ban the appearance of conflict from our assessment of evidence and our allocation of belief. But in light of what we have learned, the ambition of doing so must be deemed unrealistic. The occurrence of outright conflict in our evidence should and does discomfit us; it prompts us to re-examine both our evidence and the assumptions that underlie our frame of discernment with a view to removing that dissonance. But this effort does not always bear

*This quotation is from p. 30 of Cohen's note in *The Journal of Philosophy*. I may be quite off the mark in my interpretation of Cohen's ideas, for I often find his meaning obscure.

**The two rules of Theorem 10.5 correspond to Shackle's postulates (4) and (9) on pp. 80 and 81 of his book. The quotation is from p. 74.

fruit – at least not quickly. And using all the evidence often means using evidence that is embarrassingly conflicting.

Furthermore, the class of consonant support functions excludes many support functions that are but mildly dissonant and betray at most a potential conflict. Among these are support functions arising from the most fundamental types of probable reasoning. Consider, for example, the separable support function obtained in the case of the burglary of the sweetshop (Example 4.1). It is not consonant, for the set of left-handers neither contains nor is contained in the set of insiders. Yet its evidence against left-handers would conflict with its evidence against insiders only if further evidence exonerated left-handed insiders.

§4. Inferential Evidence

It would be a mistake, then, to treat all evidence as consonant evidence. But specific items of evidence often can be treated as consonant, and there is at least one general type of evidence that seems well adapted to such treatment. This is inferential evidence – the evidence for a cause that is provided by an effect.

Evidence is inferential with respect to a frame of discernment Θ when the possibilities in Θ are construed as possible causes of the effect to which the evidence attests. Intuitively, such evidence points toward the elements of Θ that one believes to have a strong tendency to produce the effect. It renders most plausible the elements of Θ that have the greatest such tendency, and it casts the most doubt on those that have the least such tendency. And its overall intuitive impact seems to be well described by the nested focal elements of a consonant support function.

Suppose, indeed, that $\Theta = \{\theta_1, \cdots, \theta_n\}$, the θ_i being ordered so that θ_{i+1} has a lesser tendency than θ_i to produce the effect, $i = 1, \cdots, n-1$. Then θ_1, having the greatest tendency to produce the effect, will have some positive support, and hence $\{\theta_1\}$ will be a focal element. Since it does not match θ_1 in its tendency to produce the effect, θ_2 will not deserve positive support, but the set $\{\theta_1, \theta_2\}$ will seem to deserve more support than the singleton $\{\theta_1\}$ alone, especially if θ_2 is near to θ_1 and far from θ_3 in its tendency to produce the effect; hence $\{\theta_1, \theta_2\}$ will also be a focal element. In all, we will have the nested sequence

$$\{\theta_1\} \subset \{\theta_1, \theta_2\} \subset \{\theta_1, \theta_2, \theta_3\} \subset \cdots \subset \{\theta_1, \cdots, \theta_{n-1}\} \subset \Theta$$

of focal elements. We will have, in other words, a consonant support function with contour function $f : \Theta \to [0,1]$ that obeys $f(\theta_i) < f(\theta_{i+1})$ for $i = 1, \cdots, n-1$. (This discussion assumes that no two of the θ_i have exactly the same tendency to produce the effect. If two elements θ_i and θ_{i+1} do have exactly the same such tendency, then we will set $f(\theta_i) = f(\theta_{i+1})$ and thus delete the focal element $\{\theta_1, \cdots, \theta_i\}$.)

EXAMPLE 10.1. *Refusal to Testify.* In the United States, a criminal defendant's right not to testify in his own behalf is protected by a prohibition against any adverse comment in the courtroom about a defendant's refusal to testify. Nonetheless, it is believed that juries sometimes draw negative inferences from such refusals.*

The possibility of such a negative inference arises, of course, because the defendant is thought more likely to remain silent if he is guilty than if he is innocent. If we set $\Theta = \{\theta_1, \theta_2\}$, where θ_1 corresponds to the defendant's guilt and θ_2 to his innocence, then the negative inference would result in a simple support function S focused on $\{\theta_1\}$. The quantity $S(\{\theta_1\})$ would, of course, be small; standing alone, the refusal to testify could never justify conviction. Furthermore, the jury may leave the inference undrawn if they question, in the particular instance, the basic assumption that guilt would more likely cause silence.

*See pp. 143-145 of Kalven and Zeisel's *American Jury*.

The jury may, for example, entertain the possibility that the defendant's silence is meant to protect someone else, say his wife. They will then consider a frame $\Theta = \{\theta_1, \theta_2, \theta_3\}$, where θ_1 corresponds to the defendant's guilt, θ_2 to his wife's guilt, and θ_3 to the guilt of someone else. If they assume that the protection of one's wife is equally strong a motive for silence as self-protection, they will obtain a simple support function S focused on $\{\theta_1, \theta_2\}$. If they assume that it is an even stronger motive, they will obtain a support function with focal elements $\{\theta_2\}$, $\{\theta_1, \theta_2\}$, and Θ. In either case, $S(\{\theta_1\}) = 0$; there is no direct inference to the defendant's guilt. ■

Notice that this example, like the discussion preceding it, deals with a single item of inferential evidence. It is only such single items that we can expect to be consonant. Disparate items of inferential evidence may combine to form a dissonant body of evidence, and the corresponding consonant support functions will then combine to form a dissonant support function.

As Example 10.1 illustrates, a great deal of background information is required in order to assess an item of inferential evidence. To identify all the possible causes of an effect and assess the tendency of each cause to produce the effect – this is a task that would require an immense amount of evidence in theory and that must require a tremendous amount of supposition in practice. Yet it is a task whose completion is assumed in the very frame of discernment Θ itself.

The obvious and sometimes very prominent role that supposition plays in the assessment of inferential evidence is likely to disturb us, and if we can accept such evidence at all, we are likely to accord it a very inferior status. But as I will argue in detail in Chapter 12, unproven assumptions play an essential role in the creation of all frames of discernment, no matter what the nature of the evidence. There is, of course, a great variation in the importance of this role, and it is a particularly large role in many examples of inferential evidence. But its extent does not afford us a means of distinguishing inferential evidence from other types of evidence in general.

Inferential evidence is actually quite common. We are most likely to use the term to describe cases where the inferences are problematic, but there are many other cases where the inferences can be drawn with little fanfare. And in many cases, evidence that need not be construed as inferential can be so construed if one desires. After all, there is no perception so direct but that we can hypothesize many alternatives to its having been caused by what it is ostensibly a perception of, and so we can always attribute our belief that we really perceive what we perceive to an inference.

§5. Mathematical Appendix

Proof of Theorem 10.1. Let m denote Bel's basic probability assignment.

(1) implies (2). Suppose Bel is consonant. Then its focal elements may be ordered as $A_1, \cdots, A_n$ so that $A_i \subset A_j$ whenever $i < j$. Set $A_0 = \emptyset$.

Let A and B be arbitrary subsets of Θ. Set i_1 equal to the largest integer i such that $A_i \subset A$, and set i_2 equal to the largest integer i such that $A_i \subset B$. Then $A_i \subset A$ if and only if $i \le i_1$, and $A_i \subset B$ if and only if $i \le i_2$. And $A_i \subset A \cap B$ if and only if $i \le \min(i_1, i_2)$. Hence

$$\mathrm{Bel}(A \cap B) = \sum_{i=1}^{\min(i_1,i_2)} m(A_i) = \min\left(\sum_{i=1}^{i_1} m(A_i), \sum_{i=1}^{i_2} m(A_i)\right)$$

$$= \min(\mathrm{Bel}(A), \mathrm{Bel}(B)) .$$

(2) implies (3). Suppose (2) holds. Then

$$\begin{aligned}P^*(A\cup B) &= 1 - \mathrm{Bel}(\overline{A\cup B}) = 1 - \mathrm{Bel}(\bar{A}\cap\bar{B})\\ &= 1 - \min(\mathrm{Bel}(\bar{A}), \mathrm{Bel}(\bar{B}))\\ &= \max(1-\mathrm{Bel}(\bar{A}), 1-\mathrm{Bel}(\bar{B}))\\ &= \max(P^*(A), P^*(B))\end{aligned}$$

for all $A, B \subset \Theta$.

(3) implies (4). Suppose (3) holds, and let us prove (4) by induction on the cardinality of A. If $|A| = 1$, then $A = \{\theta\}$, and (4) holds vacuously. Suppose (4) holds for $|A| = n-1$, and suppose $A = \{\theta_1, \cdots, \theta_n\}$. Then by (3),

$$\begin{aligned}P^*(A) &= \max(P^*(\{\theta_1,\cdots,\theta_{n-1}\}), P^*(\{\theta_n\}))\\ &= \max\left(\max_{1\le i\le n-1} P^*(\{\theta_i\}), P^*(\{\theta_n\})\right)\\ &= \max_{1\le i\le n} P^*(\{\theta_i\})\\ &= \max_{\theta\in A} P^*(\{\theta\}) .\end{aligned}$$

(4) implies (5). Suppose (4) holds. Suppose $A = \{\theta_1, \cdots, \theta_n\} \subset \Theta$, the θ_i being ordered so that $P^*(\{\theta_i\}) \le P^*(\{\theta_j\})$ whenever $i < j$. Then by Theorem 2.6,

$$\begin{aligned}Q(A) &= \sum_{\substack{B\subset A\\ B\ne\emptyset}} (-1)^{|B|+1} P^*(B) = \sum_{\substack{I\subset\{1,\cdots,n\}\\ I\ne\emptyset}} (-1)^{|I|+1} \max_{i\in I} P^*(\{\theta_i\})\\ &= \sum_{i\in\{1,\cdots n\}} P^*(\{\theta_i\}) \sum_{I\subset\{1,\cdots,i-1\}} (-1)^{|I|}\\ &= P^*(\{\theta_1\}) = \min_{\theta\in A} P^*(\{\theta\}) .\end{aligned}$$

Since $Q(\{\theta\}) = P^*(\{\theta\})$ for all $\theta \in \Theta$, (5) follows.

(5) implies (1). Suppose (1) does not hold. Then there exist focal elements A and B of Bel, neither of which is contained in the other. Thus we can choose $\theta_1 \in A$ such that $\theta_1 \notin B$ and $\theta_2 \in B$ such that $\theta_2 \notin A$. Since $m(A) > 0$ and $m(B) > 0$, we may conclude that

$$Q(\{\theta_1, \theta_2\}) = \sum_{\substack{C \subset \Theta \\ \theta_1, \theta_2 \in C}} m(C) \geq m(A) + \sum_{\substack{C \subset \Theta \\ \theta_2 \in C}} m(C) = m(A) + Q(\{\theta_2\})$$

$$> Q(\{\theta_2\}) ,$$

and

$$Q(\{\theta_1, \theta_2\}) = \sum_{\substack{C \subset \Theta \\ \theta_1, \theta_2 \in C}} m(C) \geq m(B) + \sum_{\substack{C \subset \Theta \\ \theta_1 \in C}} m(C) = m(B) + Q(\{\theta_1\})$$

$$> Q(\{\theta_1\}) .$$

Hence (5) does not hold. So we have established that the failure of (1) implies the failure of (5); this means that (5) implies (1).

(6) implies (1). We know from Lemma 7.2 that the focal elements of the orthogonal sum of a collection of simple support functions consist of the non-empty intersections of the foci of those simple support functions. When the foci are nested, the intersections include only the foci themselves, and (1) follows.

(1) implies (6). Suppose Bel is consonant. Then its focal elements may be arranged in nested order: $A_1 \subset A_2 \subset \cdots \subset A_n$. Set

$$s_i = \frac{m(A_i)}{1 - \sum_{j=1}^{i-1} m(A_j)}$$

for $i = 1, \cdots, n-1$, and set $s_n = 1$. Then

$$m(A_i) = s_i \prod_{j=1}^{i-1} (1-s_j)$$

for $i=1,\cdots,n$. This is because

$$\prod_{j=1}^{i} (1-s_j) = \prod_{j=1}^{i} \left(1 - \frac{m(A_j)}{1 - \sum_{k=1}^{j-1} m(A_k)}\right)$$

$$= \prod_{j=1}^{i} \frac{1 - \sum_{k=1}^{j} m(A_k)}{1 - \sum_{k=1}^{j-1} m(A_k)} = 1 - \sum_{j=1}^{i} m(A_j)$$

for $i=1,\cdots,n-1$, whence

$$s_i \prod_{j=1}^{i-1} (1-s_j) = \frac{m(A_i)}{1 - \sum_{j=1}^{i-1} m(A_j)} \left(1 - \sum_{j=1}^{i-1} m(A_j)\right) = m(A_i)$$

for $i=1,\cdots,n-1$, and

$$s_n \prod_{j=1}^{n-1} (1-s_j) = 1 - \sum_{j=1}^{n-1} m(A_j) = m(A_n) .$$

For each i, $i=1,\cdots,n-1$, let S_i denote the simple support function focused on A_i with $S(A_i) = s_i$.

Now we must consider separately the case where $A_n = \Theta$ and the case where $A_n \neq \Theta$.

If $A_n = \Theta$, then set $S = S_1 \oplus \cdots \oplus S_{n-1}$. Referring to Lemma 7.2 and using the fact that the A_i are nested, we then find that the A_i are themselves the focal elements of S. According to formula (7.6) in

Lemma 7.2, S will award Θ the basic probability number

$$\prod_{i=1}^{n-1} (1-s_i) = m(A_n),$$

and it will award A_i the basic probability number

$$\sum_{\substack{I \subset \{1,\cdots,n-1\} \\ \bigcap_{j\in I} A_j = A_i}} \left(\prod_{j\in I} s_j\right)\left(\prod_{j\in \bar{I}} (1-s_j)\right)$$

$$= s_i \prod_{j=1}^{i-1} (1-s_j) \sum_{I \subset \{i+1,\cdots,n-1\}} \left(\prod_{j\in I} s_j\right)\left(\prod_{j\in\{i+1,\cdots,n-1\}-I} (1-s_j)\right)$$

$$= s_i \prod_{j=1}^{i-1} (1-s_j) = m(A_i)$$

for $i=1,\cdots,n-1$. Hence $S = \text{Bel}$.

If $A_n \neq \Theta$, then let S_n denote the simple support function focused on A_n with $S_n(A_n) = s_n = 1$, and set $S = S_1 \oplus \cdots \oplus S_n$. Referring again to Lemma 7.2, we find that the A_i are the focal elements of S, and that S awards A_i the basic probability number

$$\sum_{\substack{I \subset \{1,\cdots,n\} \\ \bigcap_{j\in I} A_j = A_i}} \left(\prod_{j\in I} s_j\right)\left(\prod_{j\in \bar{I}} (1-s_j)\right) = m(A_i)$$

for $i=1,\cdots,n$. Hence $S = \text{Bel}$. ■

Proof of Theorem 10.2. It suffices to show that those of the $A_i \cap \mathcal{C}$ that are non-empty are nested. But $\mathcal{C}$ is itself the intersection of those A_i for which $S_i(A_i) = 1$. So by Lemma 7.2, the non-empty sets of the form $A_i \cap \mathcal{C}$ are themselves the focal elements of S. Their being nested follows from the definition of consonance. ■

Proof of Theorem 10.3. Let P^* denote the upper probability function for Bel, and let $P^*(\cdot|B)$ denote the upper probability function for $\mathrm{Bel}(\cdot|B)$ when $\mathrm{Bel}(\cdot|B)$ exists – i.e., when $P^*(B) > 0$. Then the assertion of the theorem is that Bel is consonant if and only if

$$\begin{aligned} 0 &= \min(\mathrm{Bel}(A|B), \mathrm{Bel}(\bar{A}|B)) \\ &= \min(1 - P^*(\bar{A}|B), 1 - P^*(A|B)) \\ &= 1 - \max(P^*(A|B), P^*(\bar{A}|B)) , \end{aligned}$$

or

$$\begin{aligned} 1 &= \max(P^*(A|B), P^*(\bar{A}|B)) \\ &= \max\left(\frac{P^*(A\cap B)}{P^*(B)}, \frac{P^*(\bar{A}\cap B)}{P^*(B)}\right) , \end{aligned}$$

or

$$P^*(B) = \max(P^*(A\cap B), P^*(\bar{A}\cap B)) \qquad (10.1)$$

for all $A, B \subset \Theta$ such that $P^*(B) > 0$. But (10.1) necessarily holds whenever $P^*(B) = 0$. So our task comes down to showing that

$$P^*(C\cup D) = \max(P^*(C), P^*(D)) \qquad (10.2)$$

for all $C, D \subset \Theta$ if and only if (10.1) holds for all $A, B \subset \Theta$.

(i) Suppose (10.2) holds for all $C, D \subset \Theta$. Then (10.1) certainly holds for all $A, B \subset \Theta$, for $(A\cap B) \cup (\bar{A}\cap B) = B$.

(ii) Suppose (10.1) holds for all $A, B \subset \Theta$, and consider arbitrary subsets $C, D \subset \Theta$. We may assume, without loss of generality, that $P^*(C) \geq P^*(D)$. Set $A = C$ and $B = C \cup D$. Then

$$\begin{aligned} P^*(C\cup D) &= P^*(B) = \max(P^*(A\cap B), P^*(\bar{A}\cap B)) \\ &= \max(P^*(C), P^*(\bar{C}\cap D)) \\ &= P^*(C) \\ &= \max(P^*(C), P^*(D)). \ \blacksquare \end{aligned}$$

Proof of Theorem 10.4.

(i) Suppose S is a consonant support function. Then it obeys (1) and (2) by Theorem 2.1, and it obeys (3) by Theorem 10.1.

(ii) Suppose S obeys (1), (2) and (3). Notice that if $A \subset B \subset \Theta$, then

$$S(A) = S(A \cap B) = \min(S(A), S(B)) \leq S(B) .$$

And if $A_1, \cdots, A_n$ are subsets of Θ, ordered so that

$$S(A_1) \leq S(A_2) \leq \cdots \leq S(A_n) ,$$

then

$$\sum_i S(A_i) - \sum_{i<j} S(A_i \cap A_j) + - \cdots + (-1)^{n+1} S(A_1 \cap \cdots \cap A_n)$$

$$= \sum_i S(A_i) - \sum_{i \leq n-1} \binom{n-i}{1} S(A_i) + \sum_{i \leq n-2} \binom{n-i}{2} S(A_i) - + \cdots + (-1)^{n+1} S(A_1)$$

$$= \sum_i S(A_i) \left(\binom{n-i}{0} - \binom{n-i}{1} + - \cdots + (-1)^{n-i} \binom{n-i}{n-i} \right)$$

$$= S(A_n)$$

$$\leq S(A_1 \cup \cdots \cup A_n) .$$

So S obeys (1), (2), and (3) of Theorem 2.1, and is therefore a belief function. And it follows by Theorem 10.1 that S is a consonant support function. ∎

Proof of Theorem 10.5.

(i) If Bel is a consonant belief function, then $\mathrm{Bel}(\Theta) = 1$ certainly holds. And if $\mathrm{Dou} : 2^\Theta \to [0,1]$ is defined by $\mathrm{Dou}(A) = \mathrm{Bel}(\bar{A})$, then $\mathrm{Dou}(A) = 1 - P^*(A)$ for all $A \subset \Theta$, where P^* is Bel's upper probability function. So

$$\begin{aligned}\mathrm{Dou}(A) &= 1 - P^*(A)\\ &= 1 - \max_{\theta \in A} P^*(\{\theta\})\\ &= \min_{\theta \in A} (1 - P^*(\{\theta\}))\\ &= \min_{\theta \in A} \mathrm{Bel}(\overline{\{\theta\}})\\ &= \min_{\theta \in A} \mathrm{Dou}(\{\theta\})\end{aligned}$$

for every non-empty subset A of Θ. In particular,

$$\min_{\theta \in \Theta} \mathrm{Dou}(\{\theta\}) = \mathrm{Dou}(\Theta) = \mathrm{Bel}(\emptyset) = 0 ;$$

so there must exist $\theta \in \Theta$ such that $\mathrm{Dou}(\{\theta\}) = 0$.

(ii) Suppose the function $\mathrm{Bel} : 2^\Theta \to [0,1]$ satisfies $\mathrm{Bel}(\Theta) = 1$, and the function $\mathrm{Dou} : 2^\Theta \to [0,1]$ defined by $\mathrm{Dou}(A) = \mathrm{Bel}(\bar{A})$ satisfies (1) and (2). Using (1), we find that

$$\begin{aligned}\mathrm{Bel}(A \cap B) &= \mathrm{Dou}(\overline{A \cap B}) = \mathrm{Dou}(\bar{A} \cup \bar{B})\\ &= \min_{\theta \in \bar{A} \cup \bar{B}} \mathrm{Dou}(\{\theta\})\\ &= \min\left(\min_{\theta \in \bar{A}} \mathrm{Dou}(\{\theta\}), \min_{\theta \in \bar{B}} \mathrm{Dou}(\{\theta\})\right)\\ &= \min(\mathrm{Dou}(\bar{A}), \mathrm{Dou}(\bar{B}))\\ &= \min(\mathrm{Bel}(A), \mathrm{Bel}(B))\end{aligned}$$

for all $A, B \subset \Theta$. And using (1) and (2) together, we find that

$$\mathrm{Bel}(\emptyset) = \mathrm{Dou}(\Theta) = \min_{\theta \in \Theta} \mathrm{Dou}(\{\theta\}) = 0 .$$

Hence Bel satisfies (1), (2) and (3) of Theorem 10.4, whence it is a consonant support function. ∎

CHAPTER 11. STATISTICAL EVIDENCE

La statistique est une science d'observation. Les chiffres sont les instruments à l'usage des statisticiens, et la précision de ces instruments est rendue comparable au moyen des formules tirées de la théorie des chances.

ANTOINE-AUGUSTIN COURNOT (1801-1877)

Suppose our frame of discernment Θ consists of the possible values of a parameter $\boldsymbol{\theta}$. And suppose we know that a certain aleatory experiment is governed by one of a class $\{q_\theta\}_{\theta\in\Theta}$ of chance densities on a set $\mathcal{X}$; we know that if θ is the correct value of $\boldsymbol{\theta}$ then q_θ governs the experiment. Then an observed outcome x of the experiment, since it is evidence as to which of the densities q_θ is the correct one, is also evidence as to which element of Θ is the correct value of $\boldsymbol{\theta}$. The specification of the sets Θ and $\mathcal{X}$, together with the laws $\{q_\theta\}_{\theta\in\Theta}$, may be called a *statistical specification*, and the *problem of statistical estimation* associated with such a specification is this: to determine, for each $x \in \mathcal{X}$, a support function S_x over Θ such that $S_x(A)$ is the degree of support that the observed outcome x provides for the proposition that the true value of $\boldsymbol{\theta}$ is in the subset A of Θ.

This chapter adduces and illustrates a method of calculating such a support function S_x. In §1 this method is derived from two assumptions: (1) S_x should be consonant, and (2) the plausibilities of singletons under S_x should be proportional to the chances $q_\theta(x)$. And in §2 the method is presented as a convention about weights of evidence.

Once we know how to calculate a support function S_x based on a single observation x, we also know how to calculate a support function based on a sequence of physically independent observations $x_1,\cdots,x_n$: we combine the support functions $S_{x_1},\cdots,S_{x_n}$ by Dempster's rule. In §3 I discuss this method of combining observations and compare it briefly with the method of combining that depends on product chance densities.

In §4 I discuss how our treatment of statistical evidence is related to the Bayesian treatment. And in §5 I describe the method of discounting, which allows one to restrict or eliminate the influence of discrepant observations.

In §§6-8 we turn to broader issues. In §6 and §7 we ponder what it means to associate a statistical specification with a frame of discernment, investigate how the support functions based on statistical observations are affected when one refines the frame, and discuss the element of supposition that is usually involved in a statistical specification. And in §8 we discuss how the ideas of this chapter are related to other approaches to statistical estimation and to statistical inference in general.

§1. A Convention for Assessing Statistical Evidence

When we confront a statistical specification $\{q_\theta\}_{\theta\epsilon\Theta}$ with an observation x, our intuitive feeling is that as evidence x favors those elements of Θ that attribute the greater chance to x. More precisely, we feel that x renders $\theta \epsilon \Theta$ more plausible than $\theta' \epsilon \Theta$ whenever $q_\theta(x) > q_{\theta'}(x)$. It is but a short step from this intuition to the idea that x should lend plausibility to a singleton $\{\theta\} \subset \Theta$ in strict proportion to the chance that q_θ assigns x – i.e., that x should determine a plausibility function Pl_x obeying

$$Pl_x(\{\theta\}) = cq_\theta(x) \tag{11.1}$$

for all $\theta \epsilon \Theta$, where the constant c does not depend on θ. And (11.1), together with the assumption of consonance, completely determines a plausibility function $Pl_x : 2^\Theta \to [0,1]$.

THEOREM 11.1. *Suppose* $\{q_\theta\}_{\theta\epsilon\Theta}$ *is a statistical specification on* $\mathfrak{X}$. *(And suppose that for every* $x \epsilon \mathfrak{X}$ *there exists at least one element* $\theta \in \Theta$ *such that* $q_\theta(x) > 0$.) *Then for a given* $x \in \mathcal{X}$ *there exists one and only one consonant plausibility function* $Pl_x : 2^\Theta \to [0,1]$ *that obeys* $Pl_x(\{\theta\}) = cq_\theta(x)$ *for all* $\theta \in \Theta$. *In fact, the constant* c *must be given by*

$$c = \frac{1}{\max_{\theta \in \Theta} q_\theta(x)},$$

so that Pl_x's *contour function* $f : \Theta \to [0,1]$ *is given by*

$$f(\theta) = \frac{q_\theta(x)}{\max_{\theta' \in \Theta} q_{\theta'}(x)},$$

and Pl_x *itself is given by*

$$Pl_x(A) = \frac{\max_{\theta \in A} q_\theta(x)}{\max_{\theta \in \Theta} q_\theta(x)} \tag{11.2}$$

for all non-empty subsets $A \subset \Theta$.

We thus obtain a support function $S_x : 2^\Theta \to [0,1]$ given by

$$S_x(A) = 1 - Pl_x(\bar{A}) = 1 - \frac{\max_{\theta \in \bar{A}} q_\theta(x)}{\max_{\theta \in \Theta} q_\theta(x)} \tag{11.3}$$

for all proper subsets $A \subset \Theta$.

Notice that the degrees of support and plausibility given by (11.3) and (11.2) are not implied by the general theory of evidence exposited in the preceding chapters. For neither (11.1) nor the assumption of consonance is a logical consequence of that general theory. Rather, these assumptions must be regarded as *conventions* for establishing degrees of support, conventions that can be justified only by their general intuitive appeal and by their success in dealing with particular examples.

Statisticians will immediately recognize the intuitive appeal of (11.1), for it merely translates into our vocabulary an idea that has been in the statistical literature for several centuries. Early in this century, the influential statistician R. A. Fisher taught us to pay special attention to the function $\ell: \Theta \to [0, \infty)$ given by $\ell(\theta) = q_\theta(x)$, and (11.1) merely says that this function "expresses the relative plausibilities of singletons." (See §5 of Chapter 9.) Fisher called the quantity $\ell(\theta)$ the "likelihood" of θ, but some of his successors have used the term "relative plausibility,"* and most statisticians will recognize this latter term as an appropriate one.

The assumption that the degrees of support based on a single observation x should be consonant also has its intuitive appeal. Roughly speaking, this assumption ensures that a subset of Θ is not awarded a high degree of support merely because it includes some elements θ for which $q_\theta(x)$ is high; in order to attain a given level of support a subset must include all the elements θ for which $q_\theta(x)$ exceeds a certain level.

Notice, however, that the assumption of consonance applies only to the case of a single observation. If the evidence consists of several physically independent observations $x_1, \cdots, x_n$ from the specification $\{q_\theta\}_{\theta \epsilon \Theta}$, then the overall degrees of support will be given by $S = S_{x_1} \oplus \cdots \oplus S_{x_n}$, where S_{x_i} is the consonant support function with countour function

$$f_i(\theta) = \frac{q_\theta(x_i)}{\max_{\theta' \epsilon \Theta} q_{\theta'}(x_i)} .$$

And since the different observations may point in different directions, S may be dissonant.

*George Barnard is one example. See p. 38 of his article "The Use of the Likelihood Function in Statistical Practice."

EXAMPLE 11.1. *Newman's Tramcar Problem.* In §4.8 of his *Theory of Probability*, published in 1939, Harold Jeffreys wrote as follows:

> The following problem was suggested to me several years ago by Mr. M. H. A. Newman. A man travelling in a foreign country has to change trains at a junction, and goes into the town, of the existence of which he has only just heard. He has no idea of its size. The first thing that he sees is a tramcar numbered 100. What can he infer about the number of tramcars in the town? It may be assumed for the purpose that they are numbered consecutively from 1 upwards.

The unknown parameter θ in this problem is the number of tramcars in the town, and one's evidence comes from seeing the number of a tramcar encountered at random. So it is quite natural to view the problem as a problem of statistical estimation: we take both Θ and $\mathcal{X}$ to be the set of positive integers, and we set

$$q_\theta(x) = \begin{cases} \frac{1}{\theta} & \text{if the integer } x \text{ satisfies } x \leq \theta \\ 0 & \text{if the integer } x \text{ satisfies } x > \theta . \end{cases}$$

This means that if θ is the total number of tramcars, then the chance of observing a tramcar numbered x is $\frac{1}{\theta}$ if $x \leq \theta$, and zero if $x > \theta$.

Since our exposition has dealt only with the case where the frame Θ is finite, let us assume that the total number of tramcars is no greater than some very large number N, say one million. Thus both Θ and $\mathcal{X}$ will equal the set of positive integers from 1 to N.

According to our convention, the observation $x = 100$ produces a consonant support function S over Θ, with the contour function $f:\Theta \to [0,1]$ given by

$$f(\theta) = \mathrm{Pl}(\{\theta\}) = \frac{q_\theta(100)}{\max\limits_{\theta' \in \Theta} q_{\theta'}(100)}$$

$$= 100\, q_\theta(100)$$

$$= \begin{cases} \dfrac{100}{\theta} & \text{if } \theta \geq 100 \\ 0 & \text{if } \theta < 100\,. \end{cases}$$

A few values of $\mathrm{Pl}(\{\theta\})$, $S(A)$ and $\mathrm{Pl}(A)$ are given in Table 11.1.

Table 11.1

θ	$\mathrm{Pl}(\{\theta\})$	A	S(A)	Pl(A)
99	0	$\{0,\cdots,99\}$	0	0
100	1	$\{100\}$	$\frac{1}{101}$	1
150	$\frac{2}{3}$	$\{100,\cdots,149\}$	$\frac{1}{3}$	1
200	$\frac{1}{2}$	$\{100,\cdots,199\}$	$\frac{1}{2}$	1
400	$\frac{1}{4}$	$\{100,\cdots,399\}$	$\frac{3}{4}$	1
2000	$\frac{1}{20}$	$\{100,\cdots,1999\}$	$\frac{19}{20}$	1
N	$\frac{100}{N}$	$\{100,\cdots,N\}$	1	1
		$\{101,\cdots,N\}$	0	$\frac{100}{101}$

Notice that 100 is the most plausible single value for the total number of tramcars, though there is little support for there being so few. This makes the problem of guessing the true value of $\boldsymbol{\theta}$ a bit perplexing; one might be tempted to guess 100 because it is the most plausible value, but there is little support for so low an "estimate." But of course it is not necessary to guess the true value in order to summarize the evidence. One may simply say that the number of tramcars is at least 100 and is 95% certain to be less than 2000. ■

EXAMPLE 11.2. *A Possibly Biased Coin.* Consider the statistical specification given by $\mathcal{X} = \{\text{Heads, Tails}\}$, $\Theta = \{0, 1, \cdots, 10\}$, and

$$q_\theta(\text{Heads}) = \frac{\theta}{10} \quad \text{and} \quad q_\theta(\text{Tails}) = 1 - \frac{\theta}{10} .$$

In other words, consider a coin-tossing experiment in which we know only that the true chance of heads is a multiple of $\frac{1}{10}$.

According to our conventions, a toss resulting in heads will produce the support function $S_{\text{Heads}} : 2^\Theta \to [0, 1]$ given by

$$S_{\text{Heads}}(A) = 1 - \frac{\max_{\theta \in \bar{A}} q_\theta(\text{Heads})}{\max_{\theta \in \Theta} q_\theta(\text{Heads})} = 1 - \max_{\theta \in \bar{A}} \frac{\theta}{10}$$

$$= \min_{\theta \in \bar{A}} \left(1 - \frac{\theta}{10}\right) .$$

Similarly, a toss resulting in tails will produce the support function $S_{\text{Tails}} : 2^\Theta \to [0, 1]$ given by

$$S_{\text{Tails}}(A) = \min_{\theta \in \bar{A}} \frac{\theta}{10} .$$

And if a sequence of n physically independent tosses results in k heads and $n-k$ tails, then it will produce a support function S equal to the orthogonal sum of k copies of S_{Heads} and $n-k$ copies of S_{Tails}.

Table 11.2 exhibits the support function S for the case where there are two tosses: one heads and one tails. More precisely, it gives the values of S for the intervals $\{i, i+1, \cdots, j-1, j\} \subset \Theta$, where $1 \leq i \leq j \leq 9$. Notice that the evidence is rather weak; high degrees of support are accorded only to very long intervals.

Table 11.3 similarly exhibits S for the case of ten tosses: five heads and five tails. In this case, the evidence is stronger. It clearly points to values near 5; the value 5 itself has degree of support .18, and the interval {4,5,6} has degree of support .63.

Table 11.2. Values of $S(\{i,\cdots,j\})$ resulting from two observations, one heads and one tails. All values are rounded to the nearest .01.

i \ j	1	2	3	4	5	6	7	8	9
1	.02	.07	.13	.22	.33	.47	.62	.80	1.00
2		.02	.07	.13	.22	.33	.47	.62	.80
3			.02	.07	.13	.22	.33	.47	.62
4				.02	.07	.13	.22	.33	.47
5					.02	.07	.13	.22	.33
6						.02	.07	.13	.22
7							.02	.07	.13
8								.02	.07
9									.02

Table 11.3. Values of $S(\{i,\cdots,j\})$ resulting from ten observations, five heads and five tails. All values are rounded to the nearest .01.

i \ j	1	2	3	4	5	6	7	8	9
1	.00	.02	.11	.30	.57	.81	.96	1.00	1.00
2		.02	.11	.30	.57	.81	.95	1.00	1.00
3			.08	.26	.53	.77	.91	.95	.96
4				.15	.39	.63	.77	.81	.81
5					.18	.39	.53	.57	.57
6						.15	.26	.30	.30
7							.08	.11	.11
8								.02	.02
9									.00

■

The two preceding examples should amply demonstrate the intuitive appeal of our method for determining a support function based on a single statistical observation. There is, however, at least one interesting alternative method: the method proposed by Arthur Dempster in his 1968 paper "A generalization of Bayesian inference."* That method has its own intuitive appeal, but it is much more complicated than the method of this chapter and in particular does not seem to have so simple an interpretation in terms of weights of evidence.

§2. The Weights of Evidence

In the preceding section we established a convention for determining a consonant support function $S_x : 2^\Theta \to [0,1]$ based on a single observation x from a statistical specification $\{q_\theta\}_{\theta\in\Theta}$. Since a consonant support function is necessarily separable, that convention amounts to a convention about weights of evidence.

THEOREM 11.2. *Suppose* $\{q_\theta\}_{\theta\in\Theta}$ *is a statistical specification on a set* $\mathcal{X}$, *and suppose* $x \in \mathcal{X}$. *(And suppose there exists at least one element* $\theta \in \Theta$ *such that* $q_\theta(x) > 0$.)

Enumerate the elements of Θ, *putting those that attribute the greater chance to* x *first – i.e., set* $\Theta = \{\theta_1, \cdots, \theta_n\}$, *where* $q_{\theta_1}(x) \geq q_{\theta_2}(x) \geq \cdots \geq q_{\theta_n}(x)$. *And define an assessment of evidence* w_x *over* Θ *by*

$$w_x(\{\theta_1, \cdots, \theta_i\}) = \log \frac{q_{\theta_i}(x)}{q_{\theta_{i+1}}(x)} \tag{11.4}$$

for $i = 1, \cdots, n-1$; $w_x(A) = 0$ *for all other proper* $A \subset \Theta$; *and*

*See also my paper "A Theory of Statistical Evidence," where Dempster's 1968 method is compared with the method of this chapter. (In that paper the support functions obtained by the two methods are called the *simplicial support functions* and the *linear support functions*, respectively.)

$$w_x(\Theta) = \begin{cases} \infty & \text{if (11.4) is finite for all } i \leq n-1, \\ 0 & \text{otherwise.} \end{cases}$$

(If $q_{\theta_i}(x) = q_{\theta_{i+1}}(x) = 0$, *then the logarithm in (11.4) is to be taken as zero.) Let* S_x *denote the support function determined by* w_x. *Then* S_x *is identical with the consonant support function specified by (11.3) – that is,*

$$S_x(A) = 1 - \frac{\max_{\theta \in \bar{A}} q_\theta(x)}{\max_{\theta \in \Theta} q_\theta(x)}$$

for all proper subsets $A \subset \Theta$.

The intuitive content of the assessment w_x is perhaps most accessible when we think about the weight by which w_x favors a given element $\theta \epsilon \Theta$ over another element $\theta' \epsilon \Theta$.

DEFINITION. Suppose $w : 2^\Theta \to [0,1]$ is an assessment of evidence and $\theta, \theta' \epsilon \Theta$. Then

$$\sum_{\substack{A \subset \Theta \\ \theta \epsilon A \\ \theta' \notin A}} w(A)$$

is called the *weight of evidence favoring* θ *over* θ'.

Intuitively, the evidence x favors θ over θ' only if $q_\theta(x) > q_{\theta'}(x)$, and in that case the weight with which it does so is measured by

$$\log \frac{q_\theta(x)}{q_{\theta'}(x)}.$$

Thus we might ask that w_x satisfy the condition

$$\sum_{\substack{A \subset \Theta \\ \theta \in A \\ \theta' \notin A}} w_x(A) = \begin{cases} \log \dfrac{q_\theta(x)}{q_{\theta'}(x)} & \text{if } q_\theta(x) > q_{\theta'}(x) \\ 0 & \text{otherwise .} \end{cases} \tag{11.5}$$

As it turns out, w_x does satisfy this condition. Furthermore, the condition completely determines w_x.

THEOREM 11.3. *The assessment* w_x *defined in Theorem 11.2 satisfies (11.5), and it is the only assessment over* Θ *that does so.*

Notice that if Θ has only two elements, say $\Theta = \{\theta, \theta'\}$, then condition (11.4) reduces to the simple convention that was advanced in Example 4.4. Indeed, if $q_\theta(x) \geq q_{\theta'}(x)$, then (11.4) reduces to

$$w_x(\{\theta\}) = \sum_{\substack{A \subset \Theta \\ \theta \in A \\ \theta' \notin A}} w_x(A) = \log \frac{q_\theta(x)}{q_{\theta'}(x)}$$

and

$$w_x(\{\theta'\}) = \sum_{\substack{A \subset \Theta \\ \theta' \in A \\ \theta \notin A}} w_x(A) = 0 .$$

And one thus obtains a simple support function that focuses the weight $\log \dfrac{q_\theta(x)}{q_{\theta'}(x)}$ on $\{\theta\}$.

§3. Epistemic vs. Aleatory Combination

The method of assessing statistical evidence that we learned in §1 above included a method of combining observations: the support functions based on physically independent observations were combined by Dempster's rule to obtain a support function based on all the observations together.

There is, however, an alternative. Instead of combining at the *epistemic level* by using Dempster's rule of combination, we may combine at the *aleatory level* by using product chance densities.

As we learned in §3 of Chapter 1, a sequence of n physically independent trials of an experiment governed by a chance density $q:\mathfrak{X} \to [0,1]$ can be regarded as a single trial of a compound experiment governed by the product chance density $q^n:\mathfrak{X}^n \to [0,1]$. It follows that a set of n physically independent observations from a statistical specification $\{q_\theta\}_{\theta\in\Theta}$ on $\mathfrak{X}$ may be regarded as a single observation from the *product specification* $\{q_\theta^n\}_{\theta\in\Theta}$ on $\mathfrak{X}^n$. And such a single observation can obviously be used to determine a support function over Θ.

So when we have observations $x_1,\cdots,x_n$ from a statistical specification $\{q_\theta\}_{\theta\in\Theta}$, we have a choice. We can calculate the support functions $S_{x_1},\cdots,S_{x_n}$ over Θ according to the method of §1 and then calculate the orthogonal sum $S = S_{x_1}\oplus\cdots\oplus S_{x_n}$. Or we can think of $x = (x_1,\cdots,x_n)$ as a single observation from the specification $\{q_\theta^n\}_{\theta\in\Theta}$ and use the method of §1 directly to calculate a support function S_x over Θ. This choice will make a difference; in general, S and S_x will be different.

EXAMPLE 11.3. *A Product Specification.* In Example 4.4 we considered a statistical specification consisting of two chance densities q_{θ_1} and q_{θ_2} on $\mathfrak{X} = \{\text{Heads, Tails}\}$, where

$$q_{\theta_1}(\text{Heads}) = .9, \qquad q_{\theta_1}(\text{Tails}) = .1 ,$$

and

$$q_{\theta_2}(\text{Heads}) = .3, \qquad q_{\theta_2}(\text{Tails}) = .7 .$$

The statistical evidence consisted of outcomes x_1 and x_2 from two physically independent trials: x_1 = Heads and x_2 = Tails. Thus S_{x_1} was the simple support function focused on $\{\theta_1\}$ with $S_{x_1}(\{\theta_1\}) = \frac{2}{3}$, S_{x_2} was the simple support function focused on

$\{\theta_2\}$ with $S_{x_2}(\{\theta_2\}) = \frac{6}{7}$, and $S = S_{x_1} \oplus S_{x_2}$ had the values given in Table 11.4.

Now suppose we change the specification to the corresponding product specification for pairs of trials. This specification has the set of outcomes

$$\mathcal{X}^2 = \{(\text{Heads}, \text{Heads}), (\text{Heads}, \text{Tails}), (\text{Tails}, \text{Heads}), (\text{Tails}, \text{Tails})\}$$

and chance densities $q^2_{\theta_1}$ and $q^2_{\theta_2}$, where

$$q^2_{\theta_1}((\text{Heads}, \text{Heads})) = .81, \qquad q^2_{\theta_2}((\text{Heads}, \text{Heads})) = .09,$$

$$q^2_{\theta_1}((\text{Heads}, \text{Tails})) = .09, \qquad q^2_{\theta_2}((\text{Heads}, \text{Tails})) = .21,$$

$$q^2_{\theta_1}((\text{Tails}, \text{Heads})) = .09, \qquad q^2_{\theta_2}((\text{Tails}, \text{Heads})) = .21,$$

$$q^2_{\theta_1}((\text{Tails}, \text{Tails})) = .01, \qquad q^2_{\theta_2}((\text{Tails}, \text{Tails})) = .49.$$

And the observation $x = (x_1, x_2) = (\text{Heads}, \text{Tails})$ from this specification results in a weight of evidence of

$$\log \frac{q^2_{\theta_2}((\text{Heads}, \text{Tails}))}{q^2_{\theta_1}((\text{Heads}, \text{Tails}))} = \log \frac{.21}{.09} = \log \frac{7}{3}$$

in favor of $\{\theta_2\}$ and thus in a simple support function S_x focused on $\{\theta_2\}$ with $S_x(\{\theta_2\}) = \frac{4}{7}$.

As Table 11.4 shows, S and S_x are different. Notice, however, that Pl and Pl_x award the same *relative* plausibilities to singletons.

Table 11.4

A	S(A)	Pl(A)	S_x(A)	Pl_x(A)
$\{\theta_1\}$	$\frac{2}{9}$	$\frac{3}{9}$	0	$\frac{3}{7}$
$\{\theta_2\}$	$\frac{6}{9}$	$\frac{7}{9}$	$\frac{4}{7}$	1

■

The two methods of combination always result in the same relative plausibilities for singletons. For the relative plausibilities of singletons under S_{x_i} are expressed by $\ell_i : \Theta \to [0, \infty)$, where $\ell_i(\theta) = q_\theta(x_i)$. And it follows by Theorem 9.8 that the relative plausibilities of singletons under $S = S_{x_1} \oplus \cdots \oplus S_{x_n}$ are expressed by $\ell_1 \ell_2 \cdots \ell_n$. But the relative plausibilities of singletons under $S_{\mathbf{x}}$ are expressed by $\ell : \Theta \to [0, \infty)$, where

$$\ell(\theta) = q_\theta^n((x_1, \cdots, x_n)) = q_\theta(x_1) \cdots q_\theta(x_n) = \ell_1(\theta) \cdots \ell_n(\theta) .$$

So $\ell = \ell_1 \cdots \ell_n$, and S and $S_{\mathbf{x}}$ do indeed award the same relative plausibilities to singletons. Notice, however, that $S_{\mathbf{x}}$, being consonant, will award one of its singletons plausibility one, while S may fail to award any of its singletons high plausibility. Thus the plausibilities of singletons under $S_{\mathbf{x}}$ may be many times the corresponding plausibilities under S.

It would be interesting to know more about how the two methods of combination differ in practice. Do they usually yield similar answers for the subsets of Θ that are of primary interest? And how do they diverge when unusual sets of observations $x_1, \cdots, x_n$ appear? Unfortunately, these questions are not easy to answer.

It is evident, however, that the contrast between the two methods is primarily a contrast in their treatment of conflict among the observations. By combining at the epistemic level we preserve information about such conflict. But by combining at the aleatory level we force our final support function to be consonant and thus suppress all evidence of any conflict. As we will see in §5 of this chapter and in §3 of Chapter 12, the epistemic method's ability to deal with conflict is of definite value.

§4. The Effect of a Bayesian Prior

The function $\ell : \Theta \to [0, 1]$ given by $\ell(\theta) = q_\theta(x)$ expresses the relative plausibilities of singletons under the support function $S_x : 2^\Theta \to [0, 1]$ that we learned to construct in §1. So by Theorem 9.7, the combination of S_x with a Bayesian prior Bel_0 results in the Bayesian belief function $Bel_1 : 2^\Theta \to [0, 1]$ given by

$$\mathrm{Bel}_1(\{\theta\}) = K\,\mathrm{Bel}_0(\{\theta\})\,q_\theta(x) \tag{11.6}$$

for all $\theta \epsilon \Theta$, where

$$K = \left(\sum_{\theta \epsilon \Theta} \mathrm{Bel}_0(\{\theta\})\,q_\theta(x)\right)^{-1}.$$

And (11.6) is the same as (1.7), the formula that we derived in §10 of Chapter 1 by the usual Bayesian approach of extending Bel_0 to a Bayesian belief function over $\Theta \times \mathcal{X}$. Thus our theory shows the introduction of a Bayesian prior to result in the usual Bayesian posterior, and our treatment of statistical estimation is a generalization of the Bayesian treatment.

It should be noted, though, that the support functions $\{S_x\}_{x \epsilon \mathcal{X}}$ cannot be derived by any method analogous to the derivation in Chapter 1 of the Bayesian posteriors. That is to say, they cannot be represented as the result of conditioning a belief function on the Cartesian product $\Theta \times \mathcal{X}$. So in one sense our treatment does not generalize the Bayesian treatment but rather takes a new point of view. (The method for determining support functions that Dempster advanced in his 1968 paper can be represented as the conditioning of a belief function over $\Theta \times \mathcal{X}$ and is thus a more faithful generalization of the Bayesian treatment. See §4.3 of my paper "A Theory of Statistical Evidence.")

The Bayesian treatment of statistical evidence also diverges from our treatment in that it is insensitive to the distinction between epistemic and aleatory combination. For a Bayesian prior renders superfluous all the aspects of a support function with which it is being combined except the relative plausibilities of singletons, and as we learned in the preceding section, both methods of combination result in the same relative plausibilities for singletons.

§5. Discounting Statistical Evidence

In this section I introduce the method of *discounting* a belief function and explain its importance in the combination of highly conflicting belief functions. Though this method is applicable to belief functions in general, it is particularly useful in the case of statistical evidence. And it provides a good example of how combination at the epistemic level allows one to use information about internal conflict in statistical evidence.

In earlier chapters we have always assumed that the uncertainties in our evidence were either set aside by the assumptions of the frame Θ or else taken into account by the support function $S: 2^{\Theta} \to [0,1]$. But it is possible to imagine a situation where a support function $S: 2^{\Theta} \to [0,1]$ is deemed inaccurate because it fails to take into account some particular uncertainty affecting the evidence as a whole, and in such a situation it would be natural to discount the degrees of support given by S. Indeed, if one had only a degree of trust of $1-\alpha$ in the evidence as a whole, where $0 \leq \alpha \leq 1$, then one might adopt α as one's *discount rate* and reduce the degree of support for each proper subset A of Θ from $S(A)$ to $(1-\alpha)S(A)$.

The method of discounting does not fit too neatly into our general theory of support functions. It is not obviously explicable in terms of weights of evidence, and it is inappropriate for any support function that already represents our best judgment – the fallibility of one's best judgment is no argument for changing it. Yet discounting is simple and can sometimes be useful. It can be applied to any belief function or support function:

THEOREM 11.4. *Suppose* $\mathrm{Bel}: 2^{\Theta} \to [0,1]$ *is a belief function, and suppose* $0 < \alpha < 1$. *Let* Bel^{α} *denote the function on* 2^{Θ} *given by* $\mathrm{Bel}^{\alpha}(\Theta) = 1$ *and*

$$\mathrm{Bel}^{\alpha}(A) = (1-\alpha)\mathrm{Bel}(A)$$

for all proper $A \subset \Theta$. *Then* Bel^{α} *is a support function.*

And it interacts in an interesting way with Dempster's rule of combination.

The obvious way to use discounting with Dempster's rule is to discount belief functions at different rates before combining them – discounting at higher rates those belief functions one particularly distrusts and whose influence one wants to reduce. In fact, though, discounting before combination has interesting effects even when a uniform discount rate is used.

THEOREM 11.5. *Suppose* $Bel_1, \cdots, Bel_n, Bel_{n+1}$ *are belief functions over a frame* Θ, *and suppose* $0 < \alpha < 1$. *Let* Bel_i^α *denote the belief function obtained by discounting* Bel_i *at the rate* α, *set*

$$Bel = Bel_1^\alpha \oplus \cdots \oplus Bel_n^\alpha$$

and

$$Bel' = Bel_1^\alpha \oplus \cdots \oplus Bel_n^\alpha \oplus Bel_{n+1}^\alpha = Bel \oplus Bel_{n+1}^\alpha ,$$

and let $\overline{Bel}$ *denote the average of the first* n Bel_i:

$$\overline{Bel}(A) = \frac{1}{n}(Bel_1(A) + \cdots + Bel_n(A))$$

for all $A \subset \Theta$.

(1) *Suppose* $Con(Bel_i, Bel_j) \geq n \log \frac{1}{\alpha}$ *whenever* $1 \leq i < j \leq n$. *Then*

$$|Bel(A) - \overline{Bel}(A)| \leq \frac{2\alpha}{n(1-\alpha)}$$

for all $A \subset \Theta$.

(2) *Suppose* $Con(Bel_i, Bel_{n+1}) \geq n \log \frac{1}{\alpha}$ *whenever* $1 \leq i \leq n$. *Then*

$$|Bel'(A) - Bel(A)| \leq 2\alpha^{n-1} e^{Con(Bel_1, \cdots, Bel_n)}$$

for all $A \subset \Theta$. (*Notice the following consequence: if*

$$Con(Bel_1, \cdots, Bel_n) \leq (n-1) \log \frac{1}{\alpha} - c ,$$

then

$$|Bel'(A) - Bel(A)| \leq 2e^{-c}$$

for all $A \subset \Theta$.)

As (1) of Theorem 11.5 tells us, discounting turns combination into averaging when all the belief functions being combined are highly conflicting and the discount rate is not too small. And as (2) tells us, discounting eliminates the influence of any single belief function that strongly conflicts with all the others, provided that the others do not conflict too much with each other and that the discount rate is neither too small nor too large. (Notice that this elimination takes place even though the offending belief function is discounted at the same rate as the others.)

These effects of discounting are intuitively attractive, especially when compared with what can happen when conflicting belief functions are combined without discounting. Conflict between different sources of evidence is internal evidence that something is wrong in our assessment of one or more of these sources, yet Dempster's rule will sometimes ignore such internal evidence and allow one or a few of the apparently strongest sources to dominate the others, so that the final belief function may approximate the strong and possibly erroneous one based on those sources alone. If all the belief functions we are combining are strongly conflicting, an average seems more reasonable. And if one of the belief functions strongly conflicts with the others while the others do not strongly conflict among themselves, then it seems better to eliminate the odd one than to allow it to dominate.

As I emphasized at the beginning of this section, it is appropriate to discount a support function $S : 2^{\Theta} \to [0,1]$ only if S has failed to take into account some particular uncertainty that affects the evidence as a whole. But the support function $S_x : 2^{\Theta} \to [0,1]$ based on an observation x from a statistical specification $\{q_\theta\}_{\theta \in \Theta}$ will often fail in precisely this way. Such a support function is based on the empirical fact of the occurrence of x, and like any other fact, this one may be known with less than certainty. We may think that we have observed x, but not be sure. Or more commonly, news of the occurrence of x may reach us only after a very fallible process of adjustment and transcription. Discounting S_x is the most natural way to account for such uncertainties.

If several observations $x_1, \cdots, x_n$ from the same statistical specification are compared, and the support function S_{x_1}, say, conflicts strongly with $S_{x_2}, \cdots, S_{x_n}$ while $S_{x_2}, \cdots, S_{x_n}$ do not conflict strongly among themselves, then x_1 is said to be a *discrepant* observation, or an *outlier*. The desirability of restricting the possible influence of discrepant observations has long been recognized by statisticians, and there is considerable interest in procedures for doing so.* Discounting seems to be such a procedure.

§6. Specifications on Compatible Frames

Observations from a given aleatory experiment can be assessed as evidence only with respect to certain frames of discernment, for not every frame Θ will meet the requirement that a unique chance density for the experiment be associated with each possibility $\theta \in \Theta$. There are cases, though, where the same statistical observations can be assessed with respect to several compatible frames, and in such cases we will want the results of the different assessments to be consistent. This section is devoted to showing that they are.

Though I sometimes say that one "associates" a statistical specification $\{q_\theta\}_{\theta \in \Theta}$ with a frame of discernment Θ, it must be understood that such a specification is really intrinsic to the frame Θ; the assumption that q_θ governs a certain aleatory experiment if θ is the correct value of the parameter $\boldsymbol{\theta}$ is on a par with all the other assumptions embodied in Θ, and hence cannot be violated by any other frame that is compatible with Θ. The following postulate should serve to make this understanding completely explicit.

---

*Statistical methods that restrict the influence of outliers have been dubbed *resistant* by John Tukey and *robust* by Peter Huber and others. (See *Robust Estimates of Location*, by Andrews, et al.) Most theoretical justifications of robust methods introduce the uncertainty about the quality of one's observations at the aleatory level, by enlarging the specification to include more chance densities. The method of discounting, on the other hand, introduces this uncertainty at the epistemic level.

POSTULATE. Suppose $\omega : 2^{\Theta} \to 2^{\Omega}$ is a refining, θ is an element of Θ, and $q : \mathcal{X} \to [0,1]$ is a chance density that might govern a certain aleatory experiment. Then saying that q governs the experiment if the correct value of the parameter associated with Θ is θ is equivalent to saying that q governs the experiment if the correct value of the parameter associated with Ω is in $\omega(\{\theta\})$.

In light of this understanding, we can see that once we know that a specification for an experiment is associated with one frame we know about any such specifications associated with compatible frames – in the case of each compatible frame we can tell whether there is a specification for the experiment associated with it, and if there is one we can tell what it is.

THEOREM 11.6. *Suppose a statistical specification* $\{q_\theta\}_{\theta \in \Theta}$ *is associated with a frame of discernment* Θ, *and suppose* Θ' *is a compatible frame of discernment. Then there is a statistical specification for the same experiment associated with* Θ' *if and only if* $q_{\theta_1} = q_{\theta_2}$ *whenever there exists an element of* Θ' *that is compatible with both* θ_1 *and* θ_2 *(i.e., whenever there exists* $\theta' \in \Theta$ *such that the proposition corresponding to* $\{\theta'\}$ *is compatible both with the proposition corresponding to* $\{\theta_1\}$ *and the proposition corresponding to* $\{\theta_2\}$*). When this criterion is met, the specification associated with* Θ *is given by associating with each element* θ' *of* Θ' *the chance density associated with those elements of* Θ *that are compatible with* θ'.

The main force of this theorem is captured by two simple corollaries: (1) If $\{q_\theta\}_{\theta \in \Theta}$ is associated with Θ and $\omega : 2^{\Theta} \to 2^{\Omega}$ is a refining, then there is a specification associated with Ω, and in fact q_θ is associated with each of the elements of $\omega(\{\theta\})$. (2) If $\{q_\theta\}_{\theta \in \Theta}$ is associated with Θ and $\omega : 2^{\Theta_0} \to 2^{\Theta}$ is a refining, then there is no specification associated with Θ_0 if there exists $\theta_0 \in \Theta_0$ and $\theta, \theta' \in \omega(\{\theta_0\})$ such that $q_\theta \neq q_{\theta'}$.

We are now in a position to prove that our method of assessing statistical evidence is not sensitive to a shift from one frame to a compatible one.

THEOREM 11.7.

(1) *Suppose* $\omega : 2^{\Theta} \to 2^{\Omega}$ *is a refining, suppose* $\mathcal{X}$ *is the set of possible outcomes of a certain aleatory experiment, and suppose there is a statistical specification for the experiment associated with* Θ. *(So by the preceding theorem, there is also one associated with* Ω.*) Let* S_x^{Θ} *and* S_x^{Ω} *denote the support functions that the outcome* $x \in \mathcal{X}$ *determines over* Θ *and* Ω *respectively, according to the method of §1. Then* S_x^{Ω} *is the vacuous extension of* S_x^{Θ}.

(2) *Suppose* Θ_1 *and* Θ_2 *are compatible frames, suppose* $\mathcal{X}$ *is the set of possible outcomes of a certain aleatory experiment, and suppose each frame has associated with it a statistical specification for the experiment. Let* S_x^1 *and* S_x^2 *denote the support functions that the outcome* $x \in \mathcal{X}$ *determines over* Θ_1 *and* Θ_2 *respectively, according to the method of §1. Then* S_x^1 *and* S_x^2 *are consistent.*

EXAMPLE 11.4. *A Closer Look at the Tramcar.* In Example 11.1, we used a frame of discernment Θ consisting of all the possibilities for the total number of tramcars, and such a frame can obviously be refined in many ways. We may, for example, raise the question of whether it is day or night when the man goes into the strange town, thus obtaining a refining $\omega : 2^{\Theta} \to 2^{\Omega}$ such that $\omega(\{\theta\})$ always consists of two elements of Ω, one corresponding to the proposition that it is day and there are θ tramcars, the other corresponding to the proposition that it is night and there are θ tramcars. The same density q_{θ} that is associated with an element $\theta \in \Theta$ would then be associated with each of the elements of $\omega(\{\theta\})$, and the observation of the tramcar numbered 100 would result in a support function S^{Ω} over Ω just as it resulted in the support function S over Θ in Example 11.1. And as Theorem 11.4 tells us, S^{Ω} will be the vacuous extension of S and will therefore record the same degrees of support for propositions concerned only with the number of tramcars. ■

This insensitivity contrasts sharply with the dependence on the particular frame Θ that appears when one uses a Bayesian analysis with uniform priors. (See §5 of Chapter 9.)

Let us conclude by noting an objection that might be raised against the postulate enunciated at the beginning of this section. It might be argued that some refinings can vitiate the validity of a chance density – that a refining $\omega : 2^{\Theta} \to 2^{\Omega}$ might be such that a density q_θ associated with an element $\theta \in \Theta$ ought not to be associated with the elements of $\omega(\{\theta\})$. Suppose, for example, that we refine the frame Θ of Example 11.1 by introducing, *into the frame of discernment itself*, the question of whether the first tramcar seen will bear an even or odd number. Then an element of Θ, say 250, will be split into two possibilities, one being that there are 250 tramcars and an even-numbered one will be seen first, the other being that there are 250 tramcars and an odd-numbered one will be seen first. We will hardly want to associate with either of these possibilities a chance density that assigns the chance $\frac{1}{250}$ to the man's seeing any given number from 1 through 250, indifferently to its being even or odd.

The answer to this objection is quite simple: A change in the frame Θ that does vitiate the specification $\{q_\theta\}_{\theta\in\Theta}$ in this way is not a refinement, for the specification $\{q_\theta\}_{\theta\in\Theta}$ is intrinsic to Θ, and to alter it is to produce a new frame that is incompatible with Θ.

§7. The Role of Supposition

My insistence that a statistical specification is intrinsic to its frame Θ is inspired by two thoughts: (1) In its role as our epistemic framework, our frame of discernment should incorporate all the assumptions and suppositions that we make in order to assess our evidence. (2) The knowledge that an experiment is governed by one of the chance densities in a class $\{q_\theta\}_{\theta\epsilon\Theta}$ is necessarily based on supposition rather than empirical evidence. Let us take a closer look at the nature of this supposition.

Almost everyone agrees that there is no way of being certain that an experiment is random. Indeed, it is commonplace to argue that there is never direct empirical evidence for randomness; the detection of regularities is evidence that an experiment is not random, but the failure to detect such regularities is not evidence that it is random. And in fact, statistical specifications are seldom defended as empirical facts; they are usually advanced on theoretical grounds or on grounds of their immediate utility, and they are usually thought to "fit" observations from an experiment well enough if they are not contradicted by them.

Statistical specifications vary, of course, in the extent to which they are justified as suppositions. It may be useful to distinguish several different cases.

(1) The specification may be indicated by a scientific theory that has been successful in dealing with phenomena much more general than the experiment at hand. This is surely the case where the specification is most strongly justified.

(2) The specification may be indicated by a scientific theory that has been somewhat successful but has recognized and relevant inadequacies. In this case, the randomness of the experiment may be strongly held, but the specification may be thought to be approximate.

(3) The specification and even the alleged randomness of the experiment may be a cloak thrown over certain factors that causally affect the outcomes of the experiment. In this case, the specification is inspired by the fact that one does not know enough about certain factors to make their explicit consideration useful, and it can be justified only by its success in dealing with the problem at hand.

Cases (1) and (2) seem to occur in quantum physics and genetics, but case (3) is frequent in statistical practice.

It should be noted that in all three cases, the chances specified by the true chance density are expected to approximate the frequencies with which the various outcomes will occur in a large sequence of trials of the experiment. In Cases (1) and (2) these frequencies seem to be fundamental features of nature, while in case (3) they are merely functions of the frequencies and strengths with which the ignored causes tend to be present.

EXAMPLE 11.5. *Fitting Linear Models.* One paradigmatic example of case (3) is the problem of fitting a linear model under the assumption of random errors. When one wishes to study the linear dependence of a quantity Y on quantities $X_1,\cdots,X_k$, one's first goal is usually to determine the quantities $\beta_0,\beta_1,\cdots,\beta_k$ in an equation

$$Y = \beta_0 + \beta_1 X_1 + \cdots + \beta_k X_k . \qquad (11.7)$$

(Here $Y, X_1,\cdots,X_k$ are quantities that take different values for different individuals. If the individuals are trees, for example, then one is asking for (11.7) to hold for each tree's values of $Y, X_1,\cdots,X_k$; the quantities $\beta_0,\beta_1,\cdots,\beta_k$, though unknown, are supposed to be the same for all the trees.) But if Y is in fact influenced by other factors besides $X_1,\cdots,X_k$, then actual measurements of the quantities Y and $X_1,\cdots,X_k$ for a number of individuals will yield data that will not fit (11.7) exactly for any values of $\beta_0,\beta_1,\cdots,\beta_k$. So one supposes that

$$Y = \beta_0 + \beta_1 X_1 + \cdots + \beta_k X_k + E ,$$

where E, the effect of the other factors, is thought of as a random error. If the factors subsumed under E can be assumed to operate independently of the values of $X_1,\cdots,X_k$, then it is natural to suppose that the random error E is governed by some chance density that does not depend on $\beta_0+\beta_1X_1+\cdots+\beta_kX_k$. And assuming that the values of $X_1,\cdots,X_k$ are fixed and known for a particular individual, that individual's value for Y will itself be random – it will be equal to the random error E plus the constant $\beta_0+\beta_1X_1+\cdots+\beta_kX_k$. This constant depends on the unknown parameter

$$\boldsymbol{\theta} = (\beta_0, \beta_1, \cdots, \beta_k) ,$$

and so the chance density governing Y will depend on θ; it may be denoted q_θ. The actual observed value of Y for the particular individual then becomes statistical evidence bearing on the statistical specification $\{q_\theta\}_{\theta\epsilon\Theta}$.* (Since the values of $X_1,\cdots,X_k$ will vary from individual to individual, the specification $\{q_\theta\}_{\theta\epsilon\Theta}$ will be different for each observation of Y. Nonetheless, each observation will produce a support function over Θ, and these will be combinable by Dempster's rule.)

Notice that the investigator has the task, in this example, of specifying the chance density which is assumed to govern E and which determines the specification $\{q_\theta\}_{\theta\epsilon\Theta}$. His task in setting up the problem is not likely, though, to be limited to this; even the brief exposition of the preceding paragraph reveals many assumptions that could be varied. There is, for example, a problem in deciding what is to be regarded as random. The typical answer is the one given: $X_1,\cdots,X_k$ are considered fixed and Y is considered random. But there are certainly different possibilities. And it may also be necessary to decide which causal factors to include among the $X_1,\cdots,X_k$ and which to subsume under E; this decision will directly affect the form of Θ. ■

*The problem of fitting linear models is most often dealt with in practice by the method of *least squares*. For a discussion of this method from the point of view of the Neyman-Pearson theory of statistical procedures, see Scheffé's *Analysis of Variance*.

The preceding example is not unusual in the role that it accords to supposition. The task of specification that typically precedes the solution of a problem of statistical estimation includes not just the selection of a class of chance densities but also a specification of the process that is to be regarded as random, a specification of the set $\mathcal{X}$ of its outcomes, and a specification of the form of Θ. All of these decisions will be made with a view towards facilitating the estimation problem while preserving a reasonable fit with one's data and general knowledge.

§8. Perspective

The general problem of statistical inference is to use observations as evidence about chances. And as R. A. Fisher pointed out over fifty years ago,* this general problem can be divided into two parts: the problem of *specification* and the problem of *estimation.* The problem of specification – the problem of providing a statistical specification that is both compatible with the observations and amenable to analysis – was discussed briefly in the preceding section and will be discussed again in §3 of Chapter 12. Our main concern, though, has been with the problem of estimation – the problem of using observations as evidence about the true value of a parameter $\boldsymbol{\theta}$ once a specification $\{q_\theta\}_{\theta \epsilon \Theta}$ has been set.

I began this chapter by formulating the problem of statistical estimation as the problem of using observations to assess degrees of support for subsets of the set Θ, and I believe that such an assessment of support should indeed be central both to theory and to practice once a specification is set. But I should remark that there are other ways of construing the phrase "statistical estimation." It can, for example, be construed to name the narrow and by itself pointless task of using the evidence to choose a particular $\theta \epsilon \Theta$ as one's guess or "estimate" for the true value of $\boldsymbol{\theta}$.

*See his 1922 article "On the Mathematical Foundations of Theoretical Statistics."

And many contemporary theorists, by analyzing this choice as a decision governed by the relative magnitudes of the possible errors, have come to see statistical estimation as part of a theory of decision rather than as part of a theory of evidence.

Interestingly enough, much of the theoretical work that has construed statistical estimation as a problem of choice or decision has turned out to be useful in practice for assessing evidence. It often happens, for example, that the evidence about a parameter θ can be adequately summarized by two "estimates": an estimate of θ and an estimate of the probable error of that estimate. And in many other cases an adequate method of assessing one's statistical evidence is provided by the Neyman-Pearson technique of *confidence sets*, a decision-theoretic technique that allows one to attach "confidence coefficients" to one or perhaps a nested collection of subsets of Θ.* Such a "confidence coefficient" is a lower bound on the proportion of the time that a subset chosen by the technique will turn out to contain the true value of one's parameter θ, and it does not purport to be a degree of support for the subset chosen in any particular case. But the confidence coefficients one actually obtains often seem intuitively accurate as degrees of support, and since they can be calculated without introducing prior degrees of belief they have been widely used in preference to Bayesian degrees of belief for summarizing what statistical evidence intuitively says.

I do not know how the degrees of support obtained from the conventions of this chapter or from other conventions about weights of evidence compare in practice with confidence coefficients. But it is clear that the question these degrees of support attempt to answer is the same question confidence intervals are used in practice to answer.

---

*The technique of confidence sets is due to Jerzy Neyman and E. S. Pearson and forms part of the general theory of statistical inference that bears their names. This theory is exposited in most statistics textbooks; the standard work is E. L. Lehmann's *Testing Statistical Hypotheses*.

§9. Mathematical Appendix

Proof of Theorem 11.1. The function Pl_x defined by (11.2) obviously is a consonant plausibility function satisfying $Pl_x(\{\theta\}) = cq_\theta(x)$ for all $\theta \epsilon \Theta$. And if Pl_x is any consonant plausibility function satisfying $Pl_x(\{\theta\}) = cq_\theta(x)$ for all $\theta \epsilon \Theta$, then

$$1 = Pl_x(\Theta) = \max_{\theta \epsilon \Theta} Pl_x(\{\theta\}) = c \max_{\theta \epsilon \Theta} q_\theta(x) ,$$

or

$$c = \frac{1}{\max_{\theta \epsilon \Theta} q_\theta(x)} . \blacksquare$$

Proof of Theorem 11.2. Since the subsets of Θ that are awarded positive weights of evidence by w_x are nested, it follows by Theorem 10.1 that S_x is consonant. And by (5.5), S_x's plausibility function Pl_x satisfies

$$Pl_x(\{\theta_i\}) = Q_x(\{\theta_i\}) = K \exp\left(- \sum_{\substack{B \subset \Theta \\ \theta_i \notin B}} w_x(B) \right)$$

$$= K \exp\left(- \sum_{k=1}^{i-1} \log \frac{q_{\theta_k}(x)}{q_{\theta_{k+1}}(x)} \right)$$

$$= \frac{K}{q_{\theta_1}(x)} q_{\theta_i}(x)$$

for all $\theta_i \epsilon \Theta$. So our theorem follows by Theorem 11.1. $\blacksquare$

Proof of Theorem 11.3. Assume again that $\Theta = \{\theta_1, \cdots, \theta_n\}$, where $q_{\theta_1}(x) \geq \cdots \geq q_{\theta_n}(x)$.

(1) Suppose $\theta = \theta_i$ and $\theta' = \theta_j$. Then

$$\sum_{\substack{A \subset \Theta \\ \theta \in A \\ \theta' \notin A}} w_x(A) = \sum_{\substack{k \\ i \leq k < j}} \frac{q_{\theta_k}(x)}{q_{\theta_{k+1}}(x)}$$

$$= \begin{cases} \log \dfrac{q_{\theta_i}(x)}{q_{\theta_j}(x)} & \text{if } i < j \\ 0 & \text{otherwise.} \end{cases}$$

Since $\log \dfrac{q_{\theta_i}(x)}{q_{\theta_j}(x)} = 0$ when $q_{\theta_i}(x) = q_{\theta_j}(x)$, this is the same as (11.5).

(2) Suppose the assessment $w : 2^{\Theta} \to [0, \infty]$ satisfies (11.5). Then we must show that $w = w_x$.

Suppose first that $1 \leq i < j \leq n$ and that A is a subset of Θ that contains θ_j but not θ_i. Then

$$w(A) \leq \sum_{\substack{B \subset \Theta \\ \theta_j \in B \\ \theta_i \notin B}} w(B) = 0 .$$

So $w(A) = 0$ for any $A \subset \Theta$ not of the form $\{\theta_1, \cdots, \theta_i\}$ for some i.

Now suppose $A = \{\theta_1, \cdots, \theta_i\}$, where $1 \leq i \leq n-1$. Then

$$w(\{\theta_1, \cdots, \theta_i\}) = \sum_{\substack{A \subset \Theta \\ \theta_i \in A \\ \theta_{i+1} \notin A}} w(A) = \log \frac{q_{\theta_i}(x)}{q_{\theta_{i+1}}(x)} .$$

So w satisfies (11.4).

Finally, consider $w(\Theta)$. By the definition of an assessment of evidence, $w(\Theta)$ must be infinite if all the $w(A)$ for proper $A \subset \Theta$ are finite, zero otherwise.

So we may conclude that $w = w_x$. ■

Proof of Theorem 11.4. Let m be the basic probability assignment for Bel, and define $m^{\alpha}: 2^{\Theta} \to [0,1]$ by

$$m^{\alpha}(\Theta) = (1-\alpha)\,m(\Theta) + \alpha$$

and

$$m^{\alpha}(A) = (1-\alpha)\,m(A)$$

for all proper subsets A of Θ. Then $m^{\alpha}(\emptyset) = 0$,

$$\sum_{B \subset \Theta} m^{\alpha}(B) = (1-\alpha) \sum_{B \subset \Theta} m(B) + \alpha = 1 = Bel^{\alpha}(\Theta)\ ,$$

and

$$\sum_{B \subset A} m^{\alpha}(B) = (1-\alpha) \sum_{B \subset A} m(B) = (1-\alpha)\,Bel\,(A) = Bel^{\alpha}(A)$$

for all proper subsets A of Θ. So m^{α} is a basic probability assignment and Bel^{α} is its belief function.

Since $m^{\alpha}(\Theta) \geq \alpha > 0$, Θ is the core of Bel^{α}. It follows by Theorem 7.1 that Bel^{α} is a support function. ■

LEMMA 11.1. *Suppose* $Bel_1, \cdots, Bel_k$ *are belief functions over a frame* Θ, *and suppose* $0 < \alpha < 1$. *Let* Bel_i^{α} *denote the belief function obtained by discounting* Bel_i *at the rate* α, *and set*

$$Bel = Bel_1^{\alpha} \oplus \cdots \oplus Bel_k^{\alpha}\ .$$

For each $I \subset \{1, \cdots, k\}$, *define* c_I *and* $Bel_I : 2^{\Theta} \to [0,1]$ *as follows. If* $I = \emptyset$, *set* $c_I = 1$ *and let* Bel_I *be the vacuous belief function. If* $I = \{i\}$, *set* $c_I = 1$ *and set* $Bel_I = Bel_i$. *And if* $I = \{i_1, \cdots, i_j\}$, *where* $j \geq 2$, *set*

$$c_I = e^{-Con(Bel_{i_1}, \cdots, Bel_{i_j})}$$

and set Bel_I *equal to* $\mathrm{Bel}_{i_1} \oplus \cdots \oplus \mathrm{Bel}_{i_j}$ *if that orthogonal sum exists and identically equal to zero otherwise. Then*

$$\mathrm{Bel}(A) = \frac{\sum_{I \subset \{1,\cdots,k\}} \alpha^{|\bar{I}|}(1-\alpha)^{|I|} c_I \mathrm{Bel}_I(A)}{\sum_{I \subset \{1,\cdots,k\}} \alpha^{|\bar{I}|}(1-\alpha)^{|I|} c_I}$$

for all $A \subset \Theta$, *where* $\bar{I} = \{1,\cdots,k\} - I$.

Proof of Lemma 11.1. For each i, $i=1,\cdots,k$, let Q_i and Q_i^α denote the commonality functions for Bel_i and Bel_i^α, respectively. And let Q denote the commonality function for Bel. Then

$$Q_i^\alpha(A) = \alpha + (1-\alpha)Q_i(A)$$

for all $A \subset \Theta$, $i=1,\cdots,k$. So

$$\begin{aligned} Q(A) &= K \prod_{i=1}^{k} Q_i^\alpha(A) \\ &= K\Big(\alpha^n + \alpha^{n-1}(1-\alpha)\sum_i Q_i(A) + \alpha^{n-2}(1-\alpha)^2 \sum_{i<j} Q_i(A)Q_j(A) \\ &\quad + \cdots + (1-\alpha)^n Q_1(A)\cdots Q_k(A)\Big) \end{aligned} \tag{11.8}$$

for all non-empty $A \subset \Theta$, where K is some positive constant.

Now let Q_I denote the commonality function for Bel_I whenever Bel_I is a belief function – i.e., whenever $c_I \neq 0$. And set Q_I identically equal to zero, say, when $c_I = 0$. Then $Q_\emptyset(A) = 1$ for all $A \subset \Theta$, $Q_{\{i\}} = Q_i$ for $i=1,\cdots,k$, and

$$\prod_{i \in I} Q_i(A) = c_I Q_I(A)$$

whenever $|I| \geq 2$ and A is a non-empty subset of Θ. So (11.8) becomes

$$Q(A) = K \sum_{I \subset \{1,\cdots,k\}} \alpha^{|\bar{I}|}(1-\alpha)^{|I|} c_I Q_I(A)$$

for all non-empty $A \subset \Theta$. By Theorem 2.4,

$$\mathrm{Bel}(A) = \sum_{B \subset \bar{A}} (-1)^{|B|} Q(B) = 1 + \sum_{\substack{B \subset \bar{A} \\ B \neq \emptyset}} (-1)^{|B|} Q(B)$$

$$= 1 + K \sum_{I \subset \{1,\cdots,k\}} \alpha^{|\bar{I}|}(1-\alpha)^{|I|} c_I \sum_{\substack{B \subset \bar{A} \\ B \neq \emptyset}} (-1)^{|B|} Q_I(B)$$

$$= 1 + K \sum_{I \subset \{1,\cdots,k\}} \alpha^{|\bar{I}|}(1-\alpha)^{|I|} c_I (\mathrm{Bel}_I(A) - 1)$$

for all $A \subset \Theta$. If we set $A = \emptyset$, then this yields

$$K = \left(\sum_{I \subset \{1,\cdots,k\}} \alpha^{|\bar{I}|}(1-\alpha)^{|I|} c_I \right)^{-1},$$

and our lemma then follows. ■

Lemma 11.2. *Suppose* $b > 0$, $b \geq a \geq 0$, $0 \leq \varepsilon_1 \leq \varepsilon$, *and* $0 \leq \varepsilon_2 \leq \varepsilon$. *Then*

$$\left| \frac{a + \varepsilon_1}{b + \varepsilon_2} - \frac{a}{b} \right| \leq \frac{\varepsilon}{b} .$$

Proof of Lemma 11.2. If $\varepsilon = 0$, the conclusion of the lemma is immediate. If $\varepsilon > 0$, then

$$\left|\frac{a+\varepsilon_1}{b+\varepsilon_2}-\frac{a}{b}\right| = \left|\frac{(a+\varepsilon_1)b-a(b+\varepsilon_2)}{b(b+\varepsilon_2)}\right| = \left|\frac{b\varepsilon_1-a\varepsilon_2}{b(b+\varepsilon_2)}\right|$$

$$= \frac{\varepsilon}{b}\left|\frac{b\frac{\varepsilon_1}{\varepsilon}-a\frac{\varepsilon_2}{\varepsilon}}{b+\varepsilon_2}\right| \le \frac{\varepsilon}{b}\left|\frac{b}{b+\varepsilon_2}\right|$$

$$\le \frac{\varepsilon}{b}\ . \ \blacksquare$$

Proof of Theorem 11.5. Let c_I and Bel_I be defined as in Lemma 11.1 for all $I \subset \{1,\cdots,n+1\}$.

(1) According to Lemma 11.1,

$$\mathrm{Bel}(A) = \frac{n\alpha^{n-1}(1-\alpha)\overline{\mathrm{Bel}}(A) + \sum_{\substack{I\subset\{1,\cdots,n\}\\ |I|\ge 2}} \alpha^{|\bar{I}|}(1-\alpha)^{|I|}c_I\,\mathrm{Bel}_I(A)}{\alpha^n + n\alpha^{n-1}(1-\alpha) + \sum_{\substack{I\subset\{1,\cdots,n\}\\ |I|\ge 2}} \alpha^{|\bar{I}|}(1-\alpha)^{|I|}c_I}$$

for all non-empty $A \subset \Theta$. Hence

$$\mathrm{Bel}(A)-\overline{\mathrm{Bel}}(A) = \frac{-\alpha^n\overline{\mathrm{Bel}}(A) + \sum_{\substack{I\subset\{1,\cdots,n\}\\ |I|\ge 2}} \alpha^{|\bar{I}|}(1-\alpha)^{|I|}(\mathrm{Bel}_I(A)-\overline{\mathrm{Bel}}(A))}{\alpha^n + n\alpha^{n-1}(1-\alpha) + \sum_{\substack{I\subset\{1,\cdots,n\}\\ |I|\ge 2}} \alpha^{|\bar{I}|}(1-\alpha)^{|I|}c_I}$$

and

$$|\mathrm{Bel}(A)-\overline{\mathrm{Bel}}(A)| \le \frac{\alpha^n + \sum_{\substack{I\subset\{1,\cdots,n\}\\ |I|\ge 2}} \alpha^{|\bar{I}|}(1-\alpha)^{|I|}c_I}{n\alpha^{n-1}(1-\alpha)}\ .$$

We are assuming that $\mathrm{Con}(\mathrm{Bel}_i, \mathrm{Bel}_j) \geq n \log \frac{1}{\alpha}$, or

$$c_{\{i,j\}} = e^{-\mathrm{Con}(\mathrm{Bel}_i, \mathrm{Bel}_j)} \leq \alpha^n$$

whenever $1 \leq i < j \leq n$. Since the weight of conflict among a set of belief functions either increases or remains the same as the set is enlarged, it follows that $c_I \leq \alpha^n$ for all $I \subset \{1, \cdots, n\}$ such that $|I| \geq 2$. Hence

$$|\mathrm{Bel}(A) - \overline{\mathrm{Bel}}(A)| \leq \frac{\alpha^n \left(1 + \sum_{\substack{I \subset \{1,\cdots,n\} \\ |I| \geq 2}} \alpha^{|\bar{I}|}(1-\alpha)^{|I|} \right)}{n\alpha^{n-1}(1-\alpha)}$$

$$\leq \frac{2\alpha^n}{n\alpha^{n-1}(1-\alpha)}$$

$$= \frac{2\alpha}{n(1-\alpha)}$$

for all $A \subset \Theta$.

(2) Set

$$s = \sum_{\substack{I \subset \{1,\cdots,n\} \\ |I| \geq 1}} \alpha^{|\bar{I}|}(1-\alpha)^{|I|} c_I$$

and

$$s' = \sum_{\substack{I \subset \{1,\cdots,n\} \\ |I| \geq 1}} \alpha^{|\bar{I}|}(1-\alpha)^{|I|} c_{I \cup \{n+1\}} ,$$

and set

$$r(A) = \sum_{\substack{I \subset \{1,\cdots,n\} \\ |I| \geq 1}} \alpha^{|\bar{I}|}(1-\alpha)^{|I|} c_I \mathrm{Bel}_I(A)$$

and

$$r'(A) = \sum_{I \subset \{1,\cdots,n\}} \alpha^{|\bar{I}|}(1-\alpha)^{|I|} c_{I\cup\{n+1\}} \mathrm{Bel}_{I\cup\{n+1\}}(A)$$

for all non-empty $A \subset \Theta$. Notice that s, s', $r(A)$ and $r'(A)$ are all between zero and one.

If $I = \{i_1, \cdots, i_j\}$ is a non-empty subset of $\{1, \cdots, n\}$, then

$$c_I = e^{-\mathrm{Con}(\mathrm{Bel}_{i_1}, \cdots, \mathrm{Bel}_{i_j})} \leq e^{-\mathrm{Con}(\mathrm{Bel}_1, \cdots, \mathrm{Bel}_n)} .$$

Hence

$$s \leq e^{-\mathrm{Con}(\mathrm{Bel}_1, \cdots, \mathrm{Bel}_n)} \sum_{\substack{I \subset \{1,\cdots,n\} \\ |I| \geq 1}} \alpha^{|\bar{I}|}(1-\alpha)^{|I|}$$

$$\leq e^{-\mathrm{Con}(\mathrm{Bel}_1, \cdots, \mathrm{Bel}_n)} .$$

Using Theorem 3.5 and the hypothesis that any weight of conflict involving Bel_{n+1} exceeds $n \log \frac{1}{\alpha}$, we find that

$$\frac{c_{I\cup\{n+1\}}}{c_I} = \frac{e^{-\mathrm{Con}(\mathrm{Bel}_{i_1}, \cdots, \mathrm{Bel}_{i_j}, \mathrm{Bel}_{n+1})}}{e^{-\mathrm{Con}(\mathrm{Bel}_{i_1}, \cdots, \mathrm{Bel}_{i_j})}} = e^{-\mathrm{Con}(\mathrm{Bel}_I, \mathrm{Bel}_{n+1})}$$

$$\leq \alpha^n$$

whenever $I = \{i_1, \cdots, i_j\}$ is a non-empty subset of $\{1, \cdots, n\}$ and $c_I > 0$. It follows that

$$s' \leq \alpha^n s \leq \alpha^n$$

and

$$r'(A) \leq \alpha^n s \leq \alpha^n$$

for all non-empty $A \subset \Theta$.

Now if $A \subset \Theta$ is non-empty, then Lemma 11.1 yields

$$\mathrm{Bel}(A) = \frac{r(A)}{\alpha^n + s}$$

and

$$\mathrm{Bel}'(A) = \frac{\alpha\, r(A) + (1-\alpha)(\alpha^n \mathrm{Bel}_{n+1}(A) + r'(A))}{\alpha(\alpha^n + s) + (1-\alpha)(\alpha^n + s')}$$

$$= \frac{r(A) + \alpha^{n-1}(1-\alpha)\mathrm{Bel}_{n+1}(A) + \frac{1-\alpha}{\alpha} r'(A)}{\alpha^n + s + \alpha^{n-1}(1-\alpha) + \frac{1-\alpha}{\alpha} s'}$$

$$= \frac{r(A) + 2\alpha^{n-1}\delta_1}{\alpha^n + s + 2\alpha^{n-1}\delta_2} ,$$

where $0 \le \delta_1 \le 1$ and $0 \le \delta_2 \le 1$. So by Lemma 11.2,

$$|\mathrm{Bel}'(A) - \mathrm{Bel}(A)| \le \frac{2\alpha^{n-1}}{\alpha^n + s} .$$

And

$$\frac{2\alpha^{n-1}}{\alpha^n + s} \le \frac{2\alpha^{n-1}}{s} \le 2\alpha^{n-1}\, e^{\mathrm{Con}(\mathrm{Bel}_1, \cdots, \mathrm{Bel}_n)} . \ \blacksquare$$

Proof of Theorem 11.6. Set $\Omega = \Theta \otimes \Theta'$, and let $\omega : 2^{\Theta} \to 2^{\Omega}$ and $\omega' : 2^{\Theta'} \to 2^{\Omega}$ denote the refinings to Ω.

Our postulate tells us that the specification $\{q_\theta\}_{\theta \in \Theta}$ for Θ implies the specification $\{q_\zeta\}_{\zeta \in \Omega}$ for Ω, where for each $\zeta \in \Omega$, q_ζ is equal to q_θ for that unique $\theta \in \Theta$ that is compatible with ζ – i.e., that $\theta \in \Theta$ for which $\zeta \in \omega(\{\theta\})$. Given the specification $\{q_\zeta\}_{\zeta \in \Omega}$ for Ω, our postulate also tells us that a specification $\{q'_{\theta'}\}_{\theta' \in \Theta'}$ will be associated with Θ' if and only if $q'_{\theta'} = q_\zeta$ whenever $\zeta \in \omega'(\{\theta'\})$.

So the condition that there be some specification for the process associated with Θ' is that $q_{\zeta_1} = q_{\zeta_2}$ whenever ζ_1 and ζ_2 are both in $\omega'(\{\theta'\})$ for the same θ'. In other words, we must have $q_{\theta_1} = q_{\theta_2}$ whenever there exist $\zeta_1, \zeta_2 \in \Omega$ such that $\zeta_1 \in \omega(\{\theta_1\})$, $\zeta_2 \in \omega(\{\theta_2\})$, and ζ_1 and ζ_2 are both in $\omega'(\{\theta'\})$ for the same θ'. But to say that such ζ_1 and ζ_2 exist is merely to say that $\omega(\{\theta_1\})$ and $\omega(\{\theta_2\})$ both have non-empty intersections with $\omega'(\{\theta'\})$ – i.e., that $\{\theta_1\}$ and $\{\theta_2\}$ are both compatible with $\{\theta'\}$. ■

Proof of Theorem 11.7.

(1) Let $\{q_\theta\}_{\theta \in \Theta}$ be the specification associated with Θ, so that $\{q_\zeta\}_{\zeta \in \Omega}$ is the specification associated with Ω, where $q_\zeta = q_\theta$ for that $\theta \in \Theta$ for which $\zeta \in \omega(\{\theta\})$.

The plausibility function Pl_x^{Θ} for S_x^{Θ} is given by

$$Pl_x^{\Theta}(A) = \frac{\max_{\theta \in A} q_\theta(x)}{\max_{\theta \in \Theta} q_\theta(x)}$$

for all $A \subset \Theta$. And the plausibility function Pl_x^{Ω} for S_x^{Ω} is given by

$$Pl_x^{\Omega}(A) = \frac{\max_{\zeta \in A} q_\zeta(x)}{\max_{\zeta \in \Omega} q_\zeta(x)} = \frac{\max_{\substack{\theta \in \Theta \\ \omega(\{\theta\}) \cap A \neq \emptyset}} q_\theta(x)}{\max_{\theta \in \Theta} q_\theta(x)}$$

$$= \frac{\max_{\theta \in \bar{\theta}(A)} q_\theta(x)}{\max_{\theta \in \Theta} q_\theta(x)}$$

$$= Pl_x^{\Theta}(\bar{\theta}(A))$$

for all $A \subset \Theta$, where $\bar{\theta}$ is the outer reduction for ω. It follows by Theorem 7.4 that S_x^{Ω} is the vacuous extension of S_x^{Θ}.

(2) If Ω is a common refining of Θ_1 and Θ_2, then there will be a statistical specification for the process associated with Ω, and the observation x will determine a statistical support function S_x^{Ω} over Ω. By (1), $S_x^1 = S_x^{\Omega}|2^{\Theta_1}$ and $S_x^2 = S_x^{\Omega}|2^{\Theta_2}$. Hence S_x^1 and S_x^2 are certainly compatible. ■

CHAPTER 12. THE DUAL NATURE OF PROBABLE REASONING

> Nous avons une impuissance de prouver invincible à tout le dogmatisme; nous avons une idée de la verité invincible à tout le pyrrhonisme.
>
> BLAISE PASCAL (1623-1662)

In the earlier chapters of this essay I have said little about the problem of creating a frame of discernment and little about the problem of choosing among different and incompatible frames. This is only natural, for while a mathematical theory can deal with precise and formal objects like frames of discernment, it can hardly deal with the vague and uncertain knowledge from which a frame of discernment is forged. Yet a few general remarks about incompatible frames and the difficulties of choosing among them may serve to delimit the proper range of application of our theory and of its notion of epistemic probability.

§1. The Effect of Assumptions

Though incompatible frames of discernment can be quite "incommensurable," they are often related in simple ways. And the simplest type of relation arises when one frame is obtained from another by the imposition of new assumptions.

A moment's reflection shows that the imposition of a new assumption on a frame of discernment amounts, in practical terms, to eliminating some of its possibilities. So to say that a frame Θ_1 can be obtained from another frame Θ_2 by imposing a new assumption comes down to saying that Θ_1 corresponds, as a set of possibilities, to a subset of the set of possibilities Θ_2. When two frames Θ_1 and Θ_2 are indeed related in this way, we may say that Θ_1 is an *abridgment* of Θ_2 and that Θ_2 is an *enlargement* of Θ_1.

Notice that the relation between an abridgment and an enlargement is quite different from the relation between a coarsening and a refinement. If Θ is a coarsening of Ω, then Θ and Ω are *compatible*; the two frames are based on the same concepts and assumptions, and thus when a subset of one corresponds to a subset of the other those two subsets correspond to exactly the same proposition. But if Θ_1 is an abridgment of Θ_2, then Θ_1 and Θ_2 are *incompatible*; Θ_1 incorporates an assumption that Θ_2 does not, and thus a given subset of Θ_1 represents a proposition with a somewhat different meaning than the proposition represented by the corresponding subset in the larger frame Θ_2.

EXAMPLE 12.1. *A Last Look at the Sweetshop.* In Example 4.1, we combined Sherlock Holmes' evidence against left-handers with his evidence against insiders using a frame of discernment Θ that consisted of four elements:

$$\Theta = \{LI, LO, RI, RO\} ;$$

LI represented the possibility that the thief was a left-handed insider, etc. This frame obviously embodies a number of assumptions that can and perhaps should be called into question. It is assumed for example that the thief is not ambidexterous and, more fundamentally, that the theft was perpetrated by a single individual acting alone.

Suppose we abandon the assumption that there was no collaboration. Then we will obtain an enlargement of Θ, a frame that includes points corresponding to those of Θ together with points corresponding to collaboration between one or more individuals from various of the four classes. If we use this enlargement, say Θ', to reassess the items of evidence previously thought to incriminate left-handers and insiders, then we will probably obtain a more complicated picture than we did when we used the frame Θ. In particular, we probably will not obtain positive support for the guilt of left-handed insiders. Of course, if other evidence is adduced that strongly supports the proposition that the theft was perpetrated by a single individual (i.e., supports the subset of Θ' corresponding to Θ), then the analysis using Θ' will reduce, for practical purposes, to the one using Θ.

The reader will find it easy to imagine other ways of introducing new possibilities and thus enlarging the frame in this example. Perhaps there was no burglary at all, the evidence of one being only in the imagination of Holmes and the shop's proprietor. Or perhaps the contents of the safe disappeared in accordance with some unknown physical law. There is no limit to the extent to which the frame can be enlarged by introducing possibilities of this sort, and by so enlarging it one could obviously destroy the force and the interaction of any conceivable evidence. ■

§2. The Need for Assumptions

It might be argued that we should always use the largest, most comprehensive frame of discernment available to us – that we should never abridge our frame of discernment by imposing on it assumptions that are not strongly supported by evidence. But in fact, there is never a largest frame of discernment. No matter how comprehensive a given frame is, we can think of possibilities that it excludes and thus construct an enlargement of it. Moreover, a larger, more comprehensive frame is not necessarily to be preferred to a smaller, tighter one. For as the example of the preceding section illustrates, it is always possible to enlarge a frame so as to reduce one's evidence to a collection of nullities.

To put the matter another way: probable reasoning invariably requires the use of unsupported assumptions. To say these assumptions are unsupported is not to say that there is no good reason for adopting them, but it does mean that the frame of discernment can be enlarged so that they appear as propositions that are not particularly well-supported by the evidence.

This point may be made more clearly if we classify the assumptions that are needed in our various frames of discernment into three categories: (1) the assumptions most fundamental to our thought and to the organization of our experience, (2) assumptions that we recognize as provisional and use only provisionally, and (3) the essential assumptions of successful scientific theories, assumptions which are initially provisional but which are supported by the success and the "inner necessity" of their theories.

The first of these categories has traditionally been the province of the branch of philosophy called epistemology. This category includes the assumption that the world is real and regular and the even more fundamental assumption that we can perceive that reality and regularity – that not all our experience is figmentary, not all our thought confused, not all our memory illusory. To abandon these assumptions would be to create a frame of discernment in which the evidence of our senses and our reason could hardly provide any support for anything – and could certainly provide no support for the propositions corresponding to the assumptions themselves.

Our second category of assumptions is not usually thought to be of philosophical interest, but it is of great practical importance. Whether we attend to everyday experience or to scientific research, we will find that most examples of probable reasoning are exploratory and provisional. Faced with evidence too scanty or disconnected to be useful unless analyzed under special assumptions, we provisionally make those assumptions, in the hope that they will lead us to new insights and new evidence – new knowledge that will either support the assumptions, refute them, or render them moot.

The third category of assumptions that I listed above is the category of scientific assumptions. There is a considerable literature that discusses how and in what sense the assumptions or "hypotheses" of our scientific theories are supported by evidence, and I do not intend to review that literature here. But I do wish to point out that whether a particular scientific assumption is well-supported depends on what frame the assumption is embedded in. On the one hand, it is always possible to choose so large a frame that the assumption cannot hope for positive support. In particular, we can usually enlarge the frame so as to include alternatives that so closely approximate the assumption as to be indistinguishable from it by any amount of empirical evidence. But on the other hand, powerful scientific theories often inspire a point of view and thus a frame of discernment from which their assumptions seem to be distinguished from closely approximating alternatives by simplicity and sheer naturalness. In such a frame those assumptions will certainly be well-supported.

EXAMPLE 12.2. *The Statistical Specification as an Unsupported Assumption.* Statistical specifications provide a good illustration of the preceding remarks about unsupported assumptions. For the typical specification can never be positively supported by observations from the experiment which it is alleged to govern. Moreover, this inability to obtain positive support stems from the same difficulty encountered by many quantitative scientific theories: the existence of closely approximating alternatives.

The notions of abridgment and enlargement apply in a straightforward way to statistical specifications. To enlarge a specification $\{q_\theta\}_{\theta \epsilon \Theta}$ is to exchange it for a specification $\{q'_{\theta'}\}_{\theta' \epsilon \Theta'}$, where all the densities in $\{q_\theta\}_{\theta \epsilon \Theta}$ are included among the densities in $\{q'_{\theta'}\}_{\theta' \epsilon \Theta'}$ – i.e., where for every $\theta \epsilon \Theta$ there exists at least one $\theta' \epsilon \Theta'$ such that $q_\theta = q'_{\theta'}$. If each possibility $\theta \epsilon \Theta$ has some significance apart from the density q_θ, then it may be necessary to say something about the corresponding significance of the new elements in the enlargement Θ'. But as for the specification itself, the situation is clear. An enlargement means the inclusion of additional chance densities, and an abridgment means the omission of some of the ones already included.

The exposition of statistical inference in this essay has been limited to finite specifications, just as the exposition has been limited in general to finite frames of discernment. In fact, however, the statistical specifications over a set $\mathcal{X}$ that are important in practice usually correspond to infinite but lower-dimensional subsets of the set $\mathcal{Q}$ of all chance densities over $\mathcal{X}$, where $\mathcal{Q}$ is thought of as a convex set in a linear space. Moreover, the enlargements and abridgments usually considered involve increasing or decreasing the dimension of the subset of $\mathcal{Q}$ corresponding to the specification. Now a lower dimensional subset of a set in a linear space has the property that each of its points is arbitrarily near to a point of the set that is not in the subset. Such nearby points are difficult to distinguish, and as chance densities they will be impossible to distinguish by observational evidence. It follows intuitively that when a specification $\{q_\theta\}_{\theta \in \Theta}$ is enlarged to a higher-dimensional specification $\{q'_{\theta'}\}_{\theta' \in \Theta'}$, no observational evidence will be able to give positive support to the lower dimensional subset of Θ' that corresponds to Θ.

Unlike a scientific theory, a statistical specification always has a very narrow range of application. Hence it can never have the kind of general and surprising success that can vindicate a scientific theory. And unless it itself derives from some broader theory or knowledge, its inability to derive positive support from the process it is alleged to govern means that it must always remain quite provisional. It is from this essentially provisional nature of most statistical specifications that we derive our general distrust of arguments based on statistical models. ∎

§3. Choosing our Frames of Discernment

So the practice of probable reasoning requires us to make assumptions not supported by the evidence. But how do we decide what assumptions to make? Bv what criteria do we make the choices we face when we construct a frame of discernment?

Like any creative act, the act of constructing a frame of discernment does not lend itself to thorough analysis. But we can pick out two considerations that influence it: (1) we want our evidence to interact in an interesting way, and (2) we do not want it to exhibit too much internal conflict.

Two items of evidence can always be said to interact, but they interact in an interesting way only if they jointly support a proposition more interesting than the propositions supported by either alone. In our detective tale, for example, the interaction between the evidence against left-handers and the evidence against insiders was interesting because it resulted in support for a proposition of exceptional interest – a proposition that identified a particular person as the thief. Since it depends on what we are interested in, any judgment as to whether our frame is successful in making our evidence interact in an interesting way is a subjective one. But since interesting interactions can always be destroyed by loosening relevant assumptions and thus enlarging our frame, it is clear that our desire for interesting interaction will incline us towards abridging or tightening our frame.

Our desire to avoid excessive internal conflict in our evidence will have precisely the opposite effect: it will incline us towards enlarging or loosening our frame. For internal conflict is itself a form of interaction – the most extreme form of it. And it too tends to increase as the frame is tightened, decrease as it is loosened.

EXAMPLE 12.3. *The Generation of Internal Conflict.* Suppose we have a frame of discernment consisting of three elements: $\Theta = \{a, b, c\}$. And suppose our evidence can be represented by two simple support functions: S_1 focused on $\{a, c\}$ with $S_1(\{a, c\}) = s_1$, and S_2 focused on $\{b, c\}$ with $S_2(\{b, c\}) = s_2$. Then we have a separable support function $S = S_1 \oplus S_2$ over Θ, and its weight of internal conflict W_S is zero.

Now suppose we tighten Θ by excluding the possibility c. This means that we replace Θ by a frame $\Theta' = \{a', b'\}$ which corresponds to the subset $\{a, b\}$ of Θ. Assessed with respect to Θ', our evidence will consist of a simple support function S'_1 focused on $\{a'\}$ with $S'_1(\{a'\}) = s_1$ and a simple support function S'_2 focused on $\{b'\}$ with $S'_2(\{b'\}) = s_2$. And our separable support function $S' = S'_1 \oplus S'_2$ over Θ' will have the weight of internal conflict

$$W_{S'} = -\log(1 - s_1 s_2) .$$

This weight of internal conflict will be large if both s_1 and s_2 are near to one. ■

Of course, the appearance of internal conflict will not always induce us to loosen our frame; we may find it more appropriate to suppose that our evidence is partly or wholly spurious. But there will certainly be occasions when the appearance of internal conflict will lead us to decide that the frame is too tight – that it excludes possibilities whose consideration would dissolve the apparent conflict.

The frame of discernment most appropriate to a particular instance of probable reasoning cannot, then, be determined by *a priori* considerations: it depends on what evidence is available. Notice, too, that we will tend to enlarge our frame as more evidence becomes available. For as we accumulate evidence we are likely to find more and more interactions that persist even in a looser frame and more and more conflicts that force us to looser frames.

Since we do find it necessary to adjust our assumptions to our circumstances, we use many different and incompatible frames of discernment in our practice of probable reasoning. In fact, we often consider many different frames more or less simultaneously. Even when our attention is sharply focused we tend to experiment to some extent at varying our assumptions, and as we shift the focus and level of generality of our attention we tend to vary these assumptions even more, sometimes going so far as to take for granted on one occasion something that might be the focus of our questioning on another. Consequently we will construct many different frames of discernment, which will vary in their success in organizing our experience and thus in their ultimate acceptance.

EXAMPLE 12.4. *The Goodness-of-Fit of a Statistical Specification.* The thought that a large weight of internal conflict in the evidence should make us consider enlarging our frame applies in particular to the case of a statistical specification. If our evidence consists of observations $x_1, \cdots, x_n$ that are thought to be from a statistical specification $\{q_\theta\}_{\theta \epsilon \Theta}$ and we find that the weight of conflict

$$t(x_1, \cdots, x_n) \equiv \text{Con}(S_{x_1}, \cdots, S_{x_n})$$

is very large, then we may be justified in rejecting the specification $\{q_\theta\}_{\theta \epsilon \Theta}$ in favor of a larger one – one which includes additional chance densities.*

How large should $t(x_1, \cdots, x_n)$ be allowed to be before it is considered too large? There can obviously be no absolute criterion – no particular number which $t(x_1, \cdots, x_n)$ should be forbidden to exceed. For the weight of conflict $t(x_1, \cdots, x_n)$ will usually increase without bound as the number n of observations increases, whether or not our experiment is governed by one of the densities $\{q_\theta\}_{\theta \epsilon \Theta}$. But by using the product chance densities $\{q_\theta^n\}_{\theta \epsilon \Theta}$ we can construct a standard for $t(x_1, \cdots, x_n)$, a standard which tells us how large it is relative to what might be expected when the specification $\{q_\theta\}_{\theta \epsilon \Theta}$ is valid.

---

*For a specific example where it seems appropriate to use the weight of internal conflict in this way, see §3.1 of Sandra West's *Upper and Lower Probability Inferences for the Logistic Function.*

Indeed,

$$\alpha_\theta(t) = \sum_{\substack{(x_1,\cdots,x_n)\in\mathcal{X}^n \\ t(x_1,\cdots,x_n)\geq t}} q_\theta^n((x_1,\cdots,x_n))$$

is the chance that $t(x_1,\cdots,x_n)$ will exceed a particular value t, under the assumption that the process of obtaining the n-tuple $(x_1,\cdots,x_n)$ is governed by the product density q_θ^n. So if we set

$$\alpha(t) = \max_{\theta\in\Theta} \alpha_\theta(t) ,$$

then $\alpha(t)$ will be the maximum chance, under the specification $\{q_\theta\}_{\theta\in\Theta}$, of obtaining a weight of internal conflict from n trials that is greater than or equal to t. And we may judge whether the actual weight of conflict $t(x_1,\cdots,x_n)$ is excessively large by asking whether the chance $\alpha(t(x_1,\cdots,x_n))$ is excessively small. If $\alpha(t(x_1,\cdots,x_n))$ is less than .01, say, then we will be justified in saying that $t(x_1,\cdots,x_n)$ is larger than it could usually be expected to be when the specification is valid.

The reader who is familiar with statistical inference will recognize this method of judging whether $t(x_1,\cdots,x_n)$ is "too large." Indeed, the standard approach to assessing a specification's *goodness-of-fit* has long been to choose some function t of the observations $x_1,\cdots,x_n$ and to apply this method to it. The particular function t that is used is called one's *test statistic*, and the quantity $\alpha(t(x_1,\cdots,x_n))$ is called the *observed significance level*. From the point of view of this standard approach, the present exposition is unusual only insofar as it specifies the weight of conflict $\text{Con}(S_{x_1},\cdots,S_{x_n})$ as our test statistic instead of leaving us with the need to choose a test statistic more arbitrarily.

The thought that we should keep our frame of discernment tight enough to make our evidence interact in an interesting way also applies to the case of a statistical specification. In fact, practical statisticians are always concerned to choose a specification that is "tight" enough to make the estimation problem profitable. Unfortunately, though, theoretical statisticians have not given a great deal of attention to this feature of the choice of a statistical specification. And I do not know how to give a theoretical account of it. ■

§4. The Role of Epistemic Probability

The present essay accords epistemic probability a role within and relative to our frame of discernment. In our view, a proposition attains its full meaning only as part of its frame, and its degree of support or epistemic probability is always assessed relative to that frame. When two incompatible frames are compared it may be possible, as we saw in §1 above, to find a close resemblance between a proposition in one of the frames and a proposition in the other, but no matter how close this resemblance is, the two propositions will be formally different — and their degrees of support may be very different indeed.

It follows from this view that not all epistemic probabilities are equally interesting or meaningful — that the status of an epistemic probability varies with the success and acceptance of the frame to which it is attached. If a proposition occurs in a well-accepted frame, alongside well-understood and accepted alternatives, then our judgment of its epistemic probability on given evidence will be meaningful even without explicit reference to the frame. But if a proposition is part of a completely artificial frame, then there will be little sense in talking about its epistemic probability.

One important object for which it may not be sensible to talk about an epistemic probability is one's frame of discernment itself. For we may not know how to embed our frame alongside other possibilities in an acceptable larger frame; it may not be possible, in other words, to reduce the "assumptions" involved in our frame to "propositions" in a natural larger frame. And in fact, it is precisely in the case of our most successful and acceptable frames that we would expect it to be impossible: it is when a frame resists enlargement that we consider it most successful.

It should be noted that the phrase "scientific theory" is used both to name objects which are best understood as frames of discernment and to name objects which are best understood as possibilities or as propositions within a well-accepted frame of discernment. The theory of quantum mechanics may be offered as an example of a grand scientific theory that functions primarily as frame of discernment and that has not itself been fitted into any larger framework; quantum mechanics pervades modern physics, but its very pervasiveness and our inability to list coherent alternatives to it makes it awkward to talk about whether it is supported by the evidence. On the other hand, many cases can be found where competing scientific "theories" appear as alternatives in a larger framework and hence can be compared in the extent to which they are supported by the evidence.

§5. Two Tasks

The practice of probable reasoning comprises, then, two complementary tasks: the task of constructing one's frame of discernment and the task of assessing the evidence within that frame of discernment. These two tasks differ in their spirit and their purpose, but they are equally necessary and worthy.

The construction of a frame of discernment is a creative act, and we choose among frames of discernment, in the first instance at least, by asking not which is truer but which is more beautiful and more useful. Yet those who pursue truth will not scorn this creative use of our reason, for they will realize that it is a prerequisite to their pursuit.

In contrast to the creation of a frame, the assessment of evidence within a frame is an objective and disciplined activity. Yet it will not be scorned as insignificant or humdrum by those who sing the praises of creativity. The translation of our vague and amorphous knowledge and experience into degrees of support within our frame of discernment can be a challenge to the reason and judgment of our astutest minds. It is precisely by our success in meeting this challenge that the beauty and utility of our frame is judged, and it is only in meeting it that we approach the truth and objectivity that we prize as much in probable reasoning as in mathematics.

BIBLIOGRAPHY

Andrews, D. F., Bickel, P. J., Hampel, F. R., Huber, P. J., Rogers, W. H., and Tukey, J. W.

1972 *Robust Estimates of Location,* Princeton.

Barnard, George

1967 "The Use of the Likelihood Function in Statistical Practice," *Proceedings of the Fifth Berkeley Symposium on Mathematical Statistics and Probability,* Vol. I, pp. 27-40.

Bayes, Thomas

1763 "An Essay Toward Solving a Problem in the Doctrine of Chances," *Philosophical Transactions of the Royal Society,* 53, pp. 370-418. Reprinted in 1958 on pp. 293-315 of Vol. 45 of *Biometrika*. Reprinted again in 1970 on pp. 131-153 of *Studies in the History of Statistics and Probability,* (E. S. Pearson and M. G. Kendall, eds.), Hafner.

Beran, Rudolf J.

1970 "Upper and Lower Risks and Minimax Procedures," *Proceedings of the Sixth Berkeley Symposium on Mathematical Statistics and Probability,* Vol. I, pp. 1-16.

Bernoulli, James

1713 *Ars Conjectandi,* Basel. Reprinted in 1968 by Culture et Civilisation, 115 Avenue Gabriel Lebon, Brussels. Part IV has been translated into English by Bing Sung and issued as Technical Report No. 2 of the Department of Statistics of Harvard University (February 12, 1966). It is available in microfiche from the Clearinghouse for Scientific and Technical Information, Washington, D. C.

Boole, George

1854 *An Investigation of the Laws of Thought,* London. Reprinted in 1958 by Dover, New York.

Carnap, Rudolf

1952 *The Continuum of Inductive Methods*, Chicago.

Choquet, Gustave

1953 "Theory of Capacities," *Annales de l'Institut Fourier*, V, pp. 131-295.

Cohen, L. Jonathan

1973 "A Note on Inductive Logic," *The Journal of Philosophy*, LXX, . pp. 27-40.

1970 *The Implications of Induction*, Methuen.

Courant, Richard

1937 *Differential and Integral Calculus*, Vol. I, (Second Edition) Interscience.

Cournot, Antoine-Augustin

1843 *Exposition de la théorie des chances et des probabilités*, Paris.

Cramér, Harald

1946 *Mathematical Methods of Statistics*, Princeton.

De Moivre, Abraham

1756 *The Doctrine of Chances*, Third Edition, London. Reprinted in 1967 by Chelsea, New York.

Dempster, Arthur P.

1967 "Upper and lower probabilities induced by a multivalued mapping," *Annals of Mathematical Statistics*, 38, pp. 325-339.

1968 "A generalization of Bayesian inference," *Journal of the Royal Statistical Society, Series B*, 30, pp. 205-247.

Duhem, Pierre

1906 *La Théorie physique: Son objet et sa structure*, Chevalier & Rivière, Paris. The second edition, published in 1914, was translated into English and published in 1954 as *The Aim and Structure of Physical Theory*, Princeton.

Finetti, Bruno de

1970 *Teoria Delle Probabilità*, Giulio Einaudi, Turin. Published in English in 1974 as *Theory of Probability*, Wiley.

Fisher, Ronald A.

1922 "On the Mathematical Foundations of Theoretical Statistics," *Philosophical Transactions of the Royal Society of London, Series A*, Vol. 222, pp. 309-368. Reprinted in R. A. Fisher, *Contributions to Mathematical Statistics*, Wiley, New York, 1950.

1956 *Statistical Methods and Scientific Inference*, Hafner.

Good, Irving J.

1950 *Probability and the Weighing of Evidence*, Hafner.

Granger, Thomas

1620 *Divine Logike*, London.

Hacking, Ian

1975 *The Emergence of Probability*, Cambridge.

Hall, Marshall

1967 *Combinatorial Theory*, Blaisdell.

Huber, Peter J.

1973 "The use of Choquet capacities in statistics," *Bulletin of the International Statistical Institute*, Vol. XLV, Book 4, pp. 181-188.

Huber, Peter J., and Strassen, Volker.

1973 "Minimax Tests and the Neyman-Pearson Lemma for Capacities," *Annals of Statistics*, 1, pp. 251-263.

Iverson, G. R., Longcor, W. H., Mosteller, F., Gilbert, J. P., and Youtz, C.

1971 "Bias and Runs in Dice Throwing and Recording: A Few Million Throws," *Psychometrika*, 36, pp. 1-19.

Jeffrey, Richard C.

1965 *The Logic of Decision*, McGraw-Hill.

Jeffreys, Harold

1939 *Theory of Probability*, Oxford.

Kalven, Harry, Jr., and Zeisel, Hans

1966 *The American Jury*, Little.

Keynes, John Maynard

1921 *A Treatise on Probability*, Macmillan.

Lambert, Johann Heinrich

1764 *Neues Organon*. Reprinted in 1965 as the first two volumes of Lambert's *Philosophische Schriften* by Georg Olms Verlagsbuchhandlung, Hildesheim.

Lehmann, Erich L.

1959 *Testing Statistical Hypotheses*, Wiley.

Lindley, Dennis V.

1965 *Introduction to Probability and Statistics*, Cambridge.

Poisson, Siméon-Denis

1837 *Recherches sur la probabilité des jugements en matière criminelle et en matière civile, précédées des règles générales du calcul des probabilités*, Paris.

Ramsey, Frank P.

1931 "Truth and Probability," in *The Foundations of Mathematics*, Harcourt, Brace and Co. Reprinted in 1964 on pp. 63-92 of *Studies in Subjective Probability* (Henry E. Kyburg, Jr., and Howard E. Smokler, eds.), Wiley.

Revuz, André

1955 "Fonctions croissantes et mesures sur les espaces topologique ordonnés," *Annales de l'Institut Fourier*, VI, pp. 187-269.

Rota, Gian-Carlo

1964 "Theory of Möbius Functions," *Zeitschrift fur Wahrscheinlichkeitstheorie und Verwandte Gebiete*, 2, pp. 340-368.

Savage, Leonard J.

1954 *The Foundations of Statistics*, Wiley. Second edition, Dover, 1972.

Scheffé, Henry

1959 *The Analysis of Variance*, Wiley.

Shackle, G. L. S.

1961 *Decision, Order and Time in Human Affairs*, Cambridge. Second edition, 1969.

Shafer, Glenn

1973 *Allocations of Probability*. Princeton doctoral dissertation, available from University Microfilms, Ann Arbor.

1975 "A Theory of Statistical Evidence," in *Foundations and Philosophy of Statistical Theories in the Physical Sciences*, Vol. II, (W. L. Harper and C. A. Hooker, eds.), Reidel.

Sikorski, Roman

1960 *Boolean Algebras*, Springer.

Takeuchi, Hitoshi; Uyeda, Seiya; and Kanamori, Hiroo

1967 *Debate About the Earth: Approach to Geophysics through Analysis of Continental Drift*, Freeman, Cooper & Co. Revised edition, 1970.

Thirlwall, Connop

1838 *History of Greece*, London.

West, Sandra

1971 *Upper and Lower Probability Inferences for the Logistic Function*. Harvard doctoral dissertation, available from University Microfilms, Ann Arbor.

INDEX

Library of Congress Cataloging in Publication Data

Shafer, Glenn, 1946-
A mathematical theory of evidence.

Bibliography: p.
Includes index.
1. Probabilities. 2. Mathematical statistics.
I. Title.
QA273.S48 519.2 75-30208
ISBN 0-691-08175-1
ISBN 0-691-10042-X pbk.

Milton Keynes UK
Ingram Content Group UK Ltd.
UKHW020820230924
448618UK00020B/177

9 780691 100425